RADIO ASTRONOMY: INSTRUMENTS AND OBSERVATIONS

RADIOASTRONOMICHESKIE INSTRUMENTY I NABLYUDENIYA

РАДИОАСТРОНОМИЧЕСКИЕ ИНСТРУМЕНТЫ И НАБЛЮДЕНИЯ

The Lebedev Physics Institute Series

Editor: Academician D. V. Skobel'tsyn
Director, P. N. Lebedev Physics Institute, Academy of Sciences of the USSR

Proceedings (Trudy) of the P. N. Lebedev Physics Institute

Volume 47

RADIO ASTRONOMY
Instruments and Observations

Edited by
Academician D. V. Skobel'tsyn
Director, P. N. Lebedev Physics Institute
Academy of Sciences of the USSR, Moscow

Translated from Russian

CONSULTANTS BUREAU
NEW YORK–LONDON
1971

The Russian text was published by Nauka Press in Moscow in 1969 for the
Academy of Sciences of the USSR as Volume 47 of the Proceedings (Trudy)
of the P. N. Lebedev Physics Institute.   The present translation is published under an agreement with Mezhdunarodnaya Kniga, the Soviet book ex-
port agency.

ISBN 978-1-4684-1577-3   ISBN 978-1-4684-1575-9 (eBook)
DOI 10.1007/978-1-4684-1575-9

Library of Congress Catalog Card Number 70-129264
SBN 306-10845-3

© 1971 Consultants Bureau, New York
Softcover reprint of the hardcover 1st edition 1971

A Division of Plenum Publishing Corporation
227 West 17th Street, New York, N. Y. 10011

United Kingdom edition published by Consultants Bureau, London
A Division of Plenum Publishing Company, Ltd.
Donington House, 30 Norfolk Street, London, W.C. 2, England

PREFACE

In this volume are collected scientific research reports written in recent years at the Radio-Astronomy Laboratory in the field of radio-astronomical apparatus and methods of observation.

Modern radiometers of high sensitivity, using masers and parametric amplifiers, have been installed at the RT-22 radio telescope of the Laboratory's Radio-Astronomy Station. Some of the papers are devoted to a description of these original radiometers and to their operating conditions.

Considerable space is given to a description of the meter-wave equipment and to the adjustment of the DKR-1000 diapason cruciform radio telescope. Two papers describe radio interferometers with radio relaying. This method of increasing the resolving power is extremely effective and it is expected that it will be further developed under laboratory conditions, primarily in the meter wave range.

Problems of radio-telescope design occupy an important place in contemporary radio astronomy and radio engineering. Three articles are devoted to this important line of work.

In 1966, an experiment was carried out at the Radio-Astronomy Laboratory on the investigation of the solar wind by a radio-astronomical method. The experiment gave substantially new data concerning the nonuniformity of the plasma around the sun; the experimental part of the work carried out is described in the collection.

V. V. Vitkevich, Director
Radio-Astronomy Laboratory
P.N. Lebedev Physics Institute

CONTENTS

A 5.2-cm NULL RADIOSPECTROMETER WITH A SYMMETRICAL METHOD OF RECEPTION

V. M. Gudnov, I. M. Goryachev,
V. A. Kolbasov, G. S. Misezhnikov,
R. L. Sorochenko, B. V. Sestroretskii,
and V. B. Shteinshleiger

Introduction

The development of radio-astronomical spectral research is very directly related to the progress of the measuring technology. As the sensitivity of radiospectrometers increases the possibility of separating weaker and weaker spectral signals, in which cosmic radiation is unquestionably rich, becomes greater.

The noise temperature of uhf radiometers was measured in thousands of degrees in the 50's and only one very strong radio line [1] (21 cm) was accepted. The radiation from this line, attributable to transitions in the hyperfine structure of the hydrogen atom, was measured as tens of degrees Kelvin.

Low-noise amplifiers appeared in the 60's and the noise temperature of radio-astronomical apparatus was reduced by about a factor of 10; this undoubtedly played a large part in the discovery of a number of new lines. A hydroxyl (OH) line was found at a wavelength of 18 cm [2] and several radio lines of excited hydrogen were found between 3 and 21 cm [3-6]. The intensity of these lines was only a matter of degrees and even tenths of a degree over the antenna temperature. There is little doubt that further improvements to spectrometers and an increase in their sensitivity will lead to the reception of a large number of new radio lines associated with the quantum transitions of various galactic elements and compounds.

Despite certain practical difficulties, the quantum paramagnetic amplifier (QPA) constitutes the most efficient preamplifier for spectral investigations at the present time and its potentialities are far from fully explored. Parametric, tunnel, and low-noise amplifiers of other types give high-sensitivity measurements of the continuous spectrum as a result of their wide bands; however, they cannot be regarded as equivalent to the QPA for spectral research.

There is great difficulty in realizing the theoretically high intensity offered by QPA. Increasing the sensitivity intensifies the effects of factors outside the radiometer. In some cases, the sensitivity of measurement may be determined by fluctuations in the radiation from the atmosphere, inaccuracies in the tracking of the source by the radiotelescope, and other similar causes. Of considerable importance is the "parasitic modulation" of the spectrometer, which causes a deflection of the recording device when no spectral-line signal is present.

1

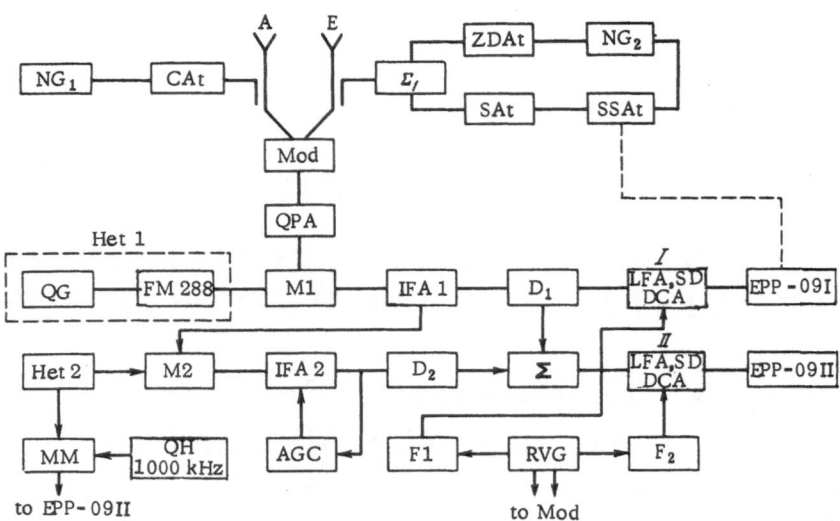

Fig. 1. Block diagram of the 5.2-cm spectral radiometer. A, antenna; E, equivalent; NG_2, compensating noise generator; SSAt, tracking system attenuator; SAt, scale attenuator; ZDAt, null displacement attenuator; Σ_1, summing slot bridge; NG_1, calibrating noise generator; CAt, calibrating noise attenuator; Mod, modulator (switch); QPA, quantum paramagnetic amplifier; M1, first mixer; IFA1, D_1, LFAI, SDI, DCAI, EPP-09I are, respectively, the intermediate-frequency amplifier, detector, low-frequency amplifier, synchronous detector, dc amplifier, and electronic potentiometer (EPP-09 type) of channel I (similarly for channel II); QG, quartz generator of first heterodyne; FM288, 288-times frequency multiplier; Het 2, tuneable second heterodyne; M2, second mixer; AGC, automatic gain control; MM, reference-mark mixer; QH 1000 kHz, quartz reference-mark heterodyne; RVG, reference voltage generator; FI and FII, phase rotators of channels I and II; Σ, subtracting device.

In this paper we shall describe a high-sensitivity radiospectrometer designed for operation at 5.2 cm and offering the possibility of separating spectral lines of very weak intensity (down to hundredths of a degree with reference to the antenna temperature). This radiometer was attached to the RT-22 radiotelescope of the Physics Institute of the Academy of Sciences in February 1966 for observing the n_{105} w n_{104} radio lines of excited hydrogen and also for measuring weak discrete sources in the continuous spectrum.

1. Block Diagram and Operating Principles of the Radiometer

The block diagram of the radiometer is shown in Fig. 1. Modulation of the external signal is effected by switching the input of the receiver between two similar reference horn irradiators taken symmetrically out of focus in the horizontal plane. As a result of this, the reception of external radiation (the radio telescope has a 10.5-min-wide polar reception pattern) takes place alternately from two directions separated by 44' from one another. The two directions of reception are completely equivalent, and the signal under consideration may be accepted through either irradiator; however, one of them is arbitrarily called the antenna (A) and the other the equivalent (E).

The radiometer employs a combination of the null method of receiving radio waves from the continuous spectrum and the differential method of separating a spectral line from a constant spectral-density noise background [7]. The radiometer has two receiving channels: I, a wide-band channel with a bandwidth ΔF and II, a narrow-band spectral-analyzer channel with a band-width δF, which may be moved along the ΔF band by adjusting the second heterodyne (Het 2). On switching over the irradiators, a signal proportional to the temperature difference of the antenna and the equivalence, $T_a - T_e$, is separated out at the output of channel I after detection in D_1. This signal is amplified at the modulation frequency (LFA) and, after synchronous detection, is used for controlling the tracking-system attenuator TSAt, which regulates the power of the compensating noise generator (NG_2). As usual in null radiometers, the system operates by equating the temperatures of the antenna and equivalent and keeping the difference $T_a - T_e$ (mean values over the band ΔF) close to zero during the investigation. At the same time the noise of channel I, detected by D_1, is fed to the subtracting device Σ, to which the noise from the spectral channel is also directed in counterphase (with respect to the modulation frequency) after detection in D_2. Provided that $\Delta F \gg \delta F$, and choosing the amplification factors of the channels in such a way that the average noise intensities in the two bands are equal

$$\overline{U_{\Delta F}^2} = \overline{U_{\delta F}^2}, \tag{1}$$

a signal at the modulation frequency with an amplitude proportional to the difference $T_a - T_e$ in the band δF is separated out at the output of the subtracting device. Subsequent amplification and recording of the spectral signal take place in LFAII, SDII, and EPP-09II in the manner usual for amplitude radiometers.

According to the analysis carried out in [7], in a system of this kind the effect of the intensity of the continuous spectrum on the output of the spectral channel will be felt only as the product of two quantities of the second order of smallness

$$\Delta T \frac{\Delta \overline{U}_{\delta F}^2}{\overline{U}_{\delta F}^2},$$

where ΔT is the error in the operation of the tracking system and $\Delta \overline{U}_{\delta F}^2$ is the drift of the noise level in the spectral channel from the balanced state represented by Eq. (1).

Thus, this radiometer includes several fundamental features which tend to reduce the influence of various interfering factors and lead to the realization of the limiting sensitivity of the apparatus under practical conditions of radio-astronomical observation. The symmetry in the antenna-equivalent system greatly reduces the effect of radiation from the atmosphere, the Earth, and various other sources of scattered radiation on the radiometer readings. Quite apart from any dependence on the position of the radiotelescope, this radiation is received almost identically by both horns and gives no component with frequency modulation.* The symmetry in switching also reduces the possibility of measuring errors due to imperfect matching, since in both positions the same uhf elements are connected to the input of the amplifying equipment.

The null method and the system of balancing the wide and narrow bands reliably protects the spectral channel from the influence of the continuous-spectrum signal. Intensity changes occurring in the course of observation, possibly as a result of errors in the tracking of the radio telescope and other causes, have little effect on the accuracy of the spectral measurements.

A result of the combination of all these factors is that, in principle, no changes take place in the radiometer with the switched input, apart from the fact that the spectrum under

---

*The limitations of this kind of system are also quite obvious, e.g., the impossibility of observing sources with angular dimensions exceeding 44' (the difference in orientation between the two receiving antennas). However, for the research purposes envisaged this limitation is not serious.

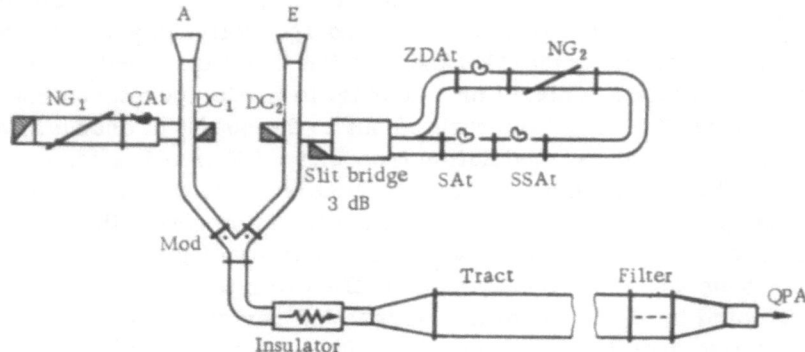

Fig. 2. Arrangement of the uhf part of the radiometer.

examination is replaced by a standard spectrum with a uniform spectral density at the input of the receiver. Only if asymmetry occurs in the manufacture of the switching elements can any undesirable "zero" drifts or parasitic effects take place. However, with the radiometer construction envisaged even this error may be eliminated in a methodical manner: by alternately observing the source along the two possible directions of reception.

Let us consider some of the structural characteristics of the various parts of the radiometer.

2. Parts of the Radiometer

Uhf Part of the Radiometer. A schematic representation of the uhf part of the radiometer is presented in Fig. 2. This incorporates horn-type irradiators, calibrating and compensating noise systems, a diode switch, a ferrite insulator, and an hf tract (waveguide). The irradiators are pyramidal horns designed for irradiation of the speculum with a fall-off at the edges corresponding to a level of 13 dB. In our case, this fall-off gives a better signal-to-noise ratio than an irradiation characteristic corresponding to a level of 10 dB.

The noise generator connected to the antenna channel (NG_1) supplies the temperature calibration of the radiometer and also serves to balance the temperature on observing the source through the equivalent irradiator. The introduction of noise from this generator takes place through a directional coupler DC_1 with an intermediate attenuation of 15 dB.

By means of the remote-controlled attenuator CAt, the calibrated noise introduced into the antenna tract may be set for two values of temperature, 15 and 31°K. Linkage of the calibrating steps is effected by the three-reading method; this involves replacing the horn-type irradiators at the radiometer input by matched loads at the temperature of liquid nitrogen, room temperature, and the temperature of boiling water. In the same way, the compensating noise signal is introduced into the equivalent channel through an analogous directional coupler DC_2. The purpose and properties of the individual units of this system will be discussed when describing the tracking system.

The radiometer switch is based on germanium-crystal diodes connecting a 120° waveguide T-junction. Depending on the sign of the voltage supplied to the diodes, the input of the receiver is connected alternately to the antenna and equivalent branches. The germanium diodes in the intermediate structure had a capacity of about 0.2 pF and a spreading resistance of under 5 Ω for a reverse bias of 4 V. The diodes were externally connected into a waveguide 48 × 24 mm in cross section along the center of the wide wall, using an inductive pin. Subsequent resonance between the inductance of the pin and the capacity of the diode ("closed" condition)

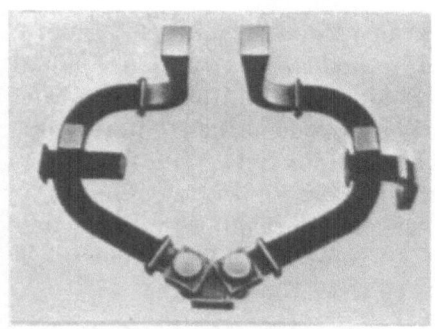

Fig. 3. Symmetrical antenna-
equivalent system.

at the working frequency was achieved by selecting the length of the loop in which the pin ended and adjusting the bias voltage. The switch has direct losses of 0.3 dB and ensures uncoupling of no less than 20 dB with the disconnected channel.

By carefully selecting pairs of diodes, the switch channels were made almost identical with respect to the attenuation introduced (difference about 0.05 dB) and the modulus and phase of the input reflection coefficient ($\Delta \Gamma \approx 0.03$, $\Delta \varphi \leq 30°$).

Under service conditions, the identity of the channels was ensured by a slight adjustment of the bias on the diodes. All the foregoing units of the radiometer are located in the focus box of the radio telescope RT-22. The general form of this part of the radiometer (with the noise sources and attenuators removed) is shown in Fig. 3. Connection between the switch and the QPA, situated in the radio telescope guide cabin [8] at a distance of about 12 m, is effected by means of a waveguide (tract) of large cross section (65 × 130 mm), which reduces the loss of hf energy.

In order to eliminate resonance phenomena in the closed space of the tract (at the working frequency some twenty undesirable types of waves may be excited) an absorbing filter for the higher-frequency waves is connected to the tract (Fig. 4). The filter has slits on its side surfaces, parallel to the electric-field vector, and longitudinal resonance slits on the upper and lower surfaces. Through these slits, the energy of all the undesirable types of waves is radiated into the space above the telescope mirror. Owing to the low noise temperature of this space, the losses associated with the radiation of the undesirable waves hardly increase the noise temperature of the tract at all.

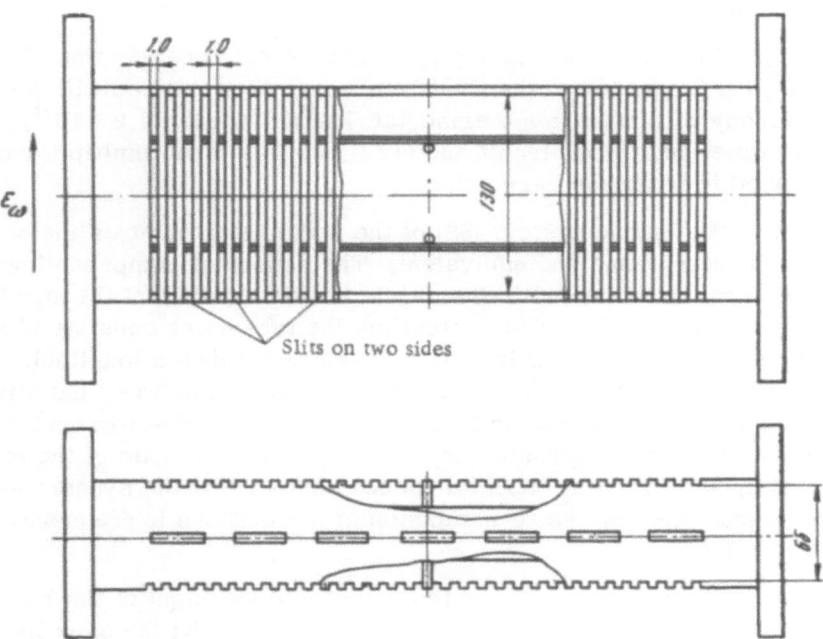

Fig. 4. Construction of the filter for absorbing undesirable types of waves.

Fig. 5. General view of the TW QPA.

For the radiation of paired (degenerate) undesirable waves of the E_{mn} and H_{mn} types, longitudinal metal strips are introduced into the filter to remove the degeneracy. In order to reduce the excitation of undesirable waves, the junctions between the 48×24 and 65×130 mm sections have a special conical-envelope shape. The waveguide tract, 12 m long, has total losses of about 0.4 dB, of which the noise contributes about 0.3 dB ($T_n \sim 20°K$).

In order to combat noise interference in the tract, which we will be discussing later, a decoupling ferrite insulator with direct losses of 0.3 dB and a decoupling factor of 44 dB had to be placed between the tract and the switch.

Quantum Paramagnetic Amplifier. A traveling-wave quantum paramagnetic amplifier (TW QPA) [9] having an amplification factor of 24 dB and a pass band (at a level of 3 dB) of 25 MHz is used in the radiometer. The noise temperature of the TW QPA together with the subsequent superheterodyne receiver was $30 \pm 5°K$. A liquid-helium cryostat of 5.5 liter capacity permitted the operation of the QPA for 11 to 13 h without additional filling.

The general form of the TW QPA is shown in Fig. 5. Practical use of the TW QPA showed it to have a very stable amplification factor and to be simple to operate. The amplifier requires practically no adjustment before operation and has only one control, i.e., for regulating the magnetic biasing current of the supplementary windings of the permanent magnet. The low sensitivity of the TW QPA to changes in input impedance (Section 3) gives it an advantage over low-noise resonator amplifiers [10].

Heterodyne 1. Since the exact tuning frequency of the receiver must be known in spectral measurements, the frequency of the first heterodyne is stabilized with a quartz crystal. The 19.8-Mc quartz heterodyne, placed in a thermostat, has a stability of $2 \cdot 10^{-6}$. This frequency is multiplied 48 times in a tube circuit and six times in a diode multiplier of the D501 type. A band filter is used to filter the harmonics.

Servo System. The servo system (SS) of the radiometer ensures the balancing of the noise temperatures of the antenna and equivalent. The source of compensating noise is a gas-discharge tube NG_2 (Fig. 2) connected to the channel of the equivalent through the directional coupler DC_2. The smooth attenuator SSAt controlling the tube noise consists of a disc coated with an absorbing film which is introduced into the waveguide through a longitudinal slit in the wide wall. By selecting the shape of the disc, a linear relationship between the attenuation and the angle of rotation is achieved. The disc is fixed to the axle of a BD-404 receiver selsyn, which is connected electrically to the transmitting selsyn fixed to the axle of the executive motor of the EPP-09 electronic potentiometer. Special checks showed that the synchronous shaft formed by these selsyns provided for the transmission of fluctuations in frequency ≤ 1.5 Hz, which corresponds to $\tau \geq 1$ sec.

The error signal proportional to $T_a - T_e$ is fed directly to the input of the EPP-09 amplifier. In this case, the executive motor of the EPP-09 will rotate until the condition $T_a = T_e$ is established in the channel of the equivalent by the smooth attenuator. Since the dc bridge of the EPP-09 is here disconnected from its input, the slide wire is used to obtain a voltage proportional to the first derivative of the error signal. This voltage is fed, in appropriate phase, to the input of the EPP-09 amplifier, which raises the stability of the servo system during transient processes.

A change in the scale of the EPP is effected by means of a discrete attenuator SAt connected in series with the SSAt. A change by the SAt is equivalent to a change in the temperature of the adjustable noise of the NG_2. The attenuator is similar in construction to the SSAt; however, its disc is fixed to the axle of the step selector providing remote control for the attenuators.

In carrying out certain scientific programs, exact measurements of small changes in signal are required. The "zero" displacement required in this case is supplied by feeding power from the NG_2 not controlled by the SSAt (power) into the channel of the equivalent. A null position convenient for observations is established by means of the discrete attenuator ZDAt, analogous in construction to the SAt. The null-displacement signal and the operating signal of the SS are combined by the 3-dB slot bridge Σ_1 and pass to the channel of the equivalent through the directional coupler DC_2.

In observing sources with an antenna temperature $T_{a.so.} < T_{NG_1}$ the null method may be used if the signal under consideration is received both by the antenna and the equivalent. All the uhf elements are wide-band, thus ensuring smoothness of the spectrum in the receiving band. An indication of the noise temperature generated by the discrete attenuators is given (in degrees Kelvin) on a light display board made of neon display tubes of the IN-1 type.

In practice, a control factor of K = 50-150 is realized in the servo system (depending on the scale). This enables the intensity of the continuous spectrum of the equivalent to be kept in close agreement with changes in the antenna spectrum. For a closed control circuit, the time constant of the integrating circuit of the radiometer is given by the expression

$$\tau_{eq} = \frac{\tau}{K},$$

where τ is the time constant of the integrating circuit with an open control circuit and K is the control factor. In view of this a value of $\tau = 300$ sec is established in the radiometer, corresponding to $\tau_{eq} = 2-6$ sec.

Spectrum Analyzer. The bandwidth of the analyzer used in investigations up to July 1966 was $\Delta F = 0.48$ MHz (for a level of 3 dB). For subsequent observations the band was narrowed to 300 kHz. The spectral-channel amplifier (IFA2), in accordance with (1), has an additional amplification factor

$$n = \sqrt{\frac{\Delta F}{\delta F}} \tag{2}$$

and is covered by the AGC circuit keeping the average intensity of the noise in the detector

$$\overline{U_{\Delta F}^2} = T_{nr}(f) Nn \tag{3}$$

constant, where N is the amplification factor with respect to the wide-band channel and $T_{nr}(f)$ is the noise temperature of the radiometer at a frequency f.

The time constant of the AGC is such that it causes no demodulation of the signal, while reacting to slow drifts in noise power. Since the noise temperature, mainly determined by the QPA, is practically constant in the analysis band, the AGC circuit actually regulates the amplification factor of the radiometer with respect to the spectral channel. This kind of stabilization gives a K_c-times reduction in both the time instabilities of the amplifying tract and the changes in the amplification factor on frequency-adjusting the amplifier ($K_c = 10$ is the feedback control factor).

Fig. 6. Low-frequency part
of the radiometer.

In order to secure an exact reading of the analyzer tuning frequency during a set of observations, frequency markers formed by the null beats of the second heterodyne frequency with the harmonics of the reference quartz generator are employed. The marks so formed (spaced at 1 MHz) are recorded on EPPII concurrently with the recording of the signal.

Low-Frequency Units of the Radiometer. Low-frequency amplification in both the wide-band and the spectral channel of the radiometer is effected by means of wide-band amplifiers, capable of passing up to the seventh harmonic of the modulation frequency. It is well known that, when compared with selective amplification at the frequency of the first harmonic, this improves the fluctuation sensitivity by a factor of $\pi/2\sqrt{2}$, or 11%. In addition to this, such amplifiers possess considerably more stable phase characteristics. In order to ease the operation of the final stages of the amplifier and the synchronous detector, the modulation frequency was reduced to 112 Hz. A multivibrator was used in the reference-voltage generator as the master generator. To make it easier to equate the half periods corresponding to the time of connecting the antenna and equivalent, the voltage from the multivibrator was divided, frequency-wise, into two by means of triggers. This voltage is fed to the synchronous detectors through fantastron delay circuits, which constitute phase rotators. The whole lf part of the radiometer is situated in the central cabin of the RT-22 radio telescope. A photograph of this part of the system is shown in Fig. 6.

3. Interference of Noise in the Tract. "Parasitic Modulation"

If the propagation time of the signal in the tract is shorter than the oscillation correlation time $1/\Delta F$, interference takes place between the noise reflected from the beginning of the tract and the unreflected noise. As a result of this the spectral density of the noise in the receiver, connected to one end of the tract, becomes dependent on the length of the tract l and the frequency f. This dependence still occurs on using the modulation method of reception. The value of the modulation component of the frequency-dependent interference term will be determined by the difference between the electrical parameters of the switching elements. According to [11], this value (expressed in degrees Kelvin) equals

$$\delta T\,(f) = 2T_{\,i} \left[\Gamma_a \cos \left(\frac{4\pi l_a}{v_p} f \right) - \Gamma_e \cos \left(\frac{4\pi l_e}{v_p} f \right) \right], \tag{4}$$

where T_i is the temperature of the interfering noise in the tract; Γ_a and Γ_e are the voltage reflection coefficients in the antenna and equivalent arms; l_a and l_e are the corresponding geometrical lengths of the arms; and v_p is the phase velocity in the tract in which the interference takes place.

Since such a modulation component leads to a deflection of the recording device of the spectrometer in the absence of any of the spectral signals, it is sometimes called "parasitic modulation." This has a very harmful effect on the sensitivity of the spectrometer and may lead to erroneous results.

It follows from Eq. (4) that a fundamental procedure in combating the parasitic modulation due to noise is the careful matching of the uhf elements. At the same time it is desirable

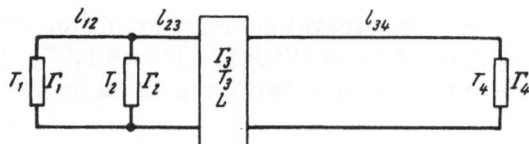

Fig. 7. Equivalent circuit of the uhf part of the radiometer. Γ_1, Γ_2, Γ_3, Γ_4 are, respectively, the reflection coefficients from the irradiator, switch, insulator, and input of the QPA; T_1 ($=T_a$), temperature of the antenna; T_2, noise temperature of the switch; T_3, temperature of the insulator; T_4, temperature of the QPA noise radiated into the tract; l_{12}, distance between the irradiator and switch; l_{23}, distance between the switch and the insulator; l_{34}, distance between the insulator and the QPA input; L, decoupling of the insulator.

to minimize the length of the tract subject to interference, since this will weaken the frequency dependence of the parasitic-modulation effect.

Let us represent the uhf part of the radiometer by the equivalent circuit of Fig. 7. We shall consider the noise of the main uhf elements, which are concentrated at the point of connection. This procedure is completely valid for elucidating interference phenomena, since the noise of the distributed elements (due to linear losses in the tract) will yield no interference.

We see from Fig. 7 that the period of the interference signal, modulated with the switching frequency, is determined by the lengths of the tracts $l_{23} + l_{12}$, $l_{34} + l_{23}$, and $l_{34} + l_{23} + l_{12}$. The most unfavorable of these are the two last sections, which are about 12 m long. The interference developing in this section has a period of $\Delta f_l = v_p / 2l \approx 13$ MHz, i.e., almost equal to the band in which the spectral analysis is being carried out.

Despite the symmetry principle employed, because of the tolerances involved in manufacture and a certain scatter in the parameters of the switching diodes, the switching elements were not really identical.

As noted earlier, by dint of tuning and selecting diodes with similar characteristics, a value of $(\Gamma_a - \Gamma_e) = 0.03$ was obtained with very similar phases of the reflection coefficients. Even in this case, however, for T ~ 10°K we find from (4) that

$$\delta T\,(f) = 0.6 \cos\left[\frac{4\pi l_a}{v_{\mathrm{p}}} f\right], \tag{5}$$

which is quite unacceptable for a 13-MHz period of interference. The interference of noise in the tract is not the only cause of parasitic modulation. Let us estimate the analogous effect due to a change in the amplification factor resulting from a difference in the impedances of the antenna and equivalent arms.

It may be shown that, using simplifying assumptions,* the relative variation in the power amplification factor (gain) G, of a TW QPA due to a change in impedance on switching, may to a first approximation be expressed by the following formula:

$$\frac{\Delta G\,(f)}{G} \leqslant 2\,[\Gamma_a - \Gamma_e]\frac{1}{\gamma}\cos\left[\frac{4\pi}{v_{\mathrm{p}}}\,(l_{34} + l_{23})\,f\right], \tag{6}$$

where $\gamma = (1/\Gamma_{\mathrm{out}'})\sqrt{L/G}$ is the "effective decoupling" of the TW QPA; Γ_{out} is the reflection coefficient of the output of the TW QPA; G is the resultant power gain in the forward direction; L is the resultant power attenuation in the backward direction between inhomogeneities (nonuniformities) with reflection factors $\Gamma_{a.e}$ and Γ_{out}.

*It is assumed that Γ_a and Γ_e differ only in amplitude (this is almost true) and that there are no other reflections in the tract.

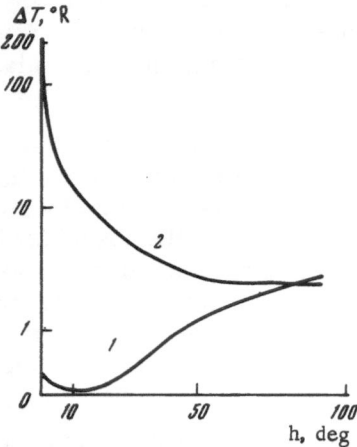

Fig. 8. Increment in the temperature background as a function of the angle of elevation of the observed point. 1) Measured temperature difference $T_a - T_e = \Delta T$ for a symmetrical system (RT-22, 5760 MHz); 2) uncompensated temperature background according to the measurements of [13] (BTL, 5650 MHz).

We note that for a traveling-wave amplifier the "effective decoupling" (in decibels) is $[20 \log 1/\Gamma_{out}]_{dB}$ larger than for a resonator amplifier with the same L/G ratio.

The change in the readings at the output of the spectral channel (in degrees Kelvin) is

$$\delta T_{QPA}(t) = \frac{\Delta G(t)}{G} T_{in}, \qquad (7)$$

where T_{in} is the temperature of the noise passing into the QPA input.

Putting $(\Gamma_a - \Gamma_e) \approx 0.03$, $\gamma = 20$ (a typical value for the QPA employed in the absence of additional external decouplings) and $T_{in} = 70°$K, we obtain

$$\delta T_{QPA}(f) \approx 0.2 \cos \left[\frac{4\pi(l_{34} + l_{23})}{v_p} f \right] °K, \qquad (8)$$

which agrees with (5) in order of magnitude and has the same period of interference.

In order to remove the parasitic modulation with a frequency period of 13 MHz due to both the interference of noise and changes in the amplification of the QPA, a ferrite insulator was placed between the switch and long tract.* With a decoupling of $L_n = 44$ dB the insulator reduces the effects defined by expressions (5) and (8) by $\sqrt{L_n} = 160$ times, and this certainly justifies a slight increase in the noise temperature of the radiometer. It is true that in this case there is interference due to the 300°K noise of the insulator itself, propagated in the direction of the antenna and reflected from the switch. However, thanks to the very short interference arm, the frequency gradient is comparatively slight. An additional reduction in this gradient may be achieved by using a circulator, with de-excitation of the radiation of the free arm, for decoupling [12].

The whole procedure relating to the reduction of interference phenomena in the tract and to the parasitic modulation was the hardest and most complicated aspect in the adjustment of the radiometer. As a result of the measures taken, the slope of the zero line in the spectrograms was reduced to between 0.004 and 0.05 deg/MHz, which was entirely satisfactory for the research undertaken.

4. Practical Parameters of the Radiometer and Some Results

The main characteristic of the radiometer is its sensitivity. In the wide-band channel ($\Delta F = 25$ MHz) a sensitivity of $\delta T = 0.04-0.05°$ was realized with $\tau = 1$ sec, which corresponds to the general noise temperature of the radiometer $T_{nr} = 130-160°$K. The noise temperature so determined agrees quite closely with the calculated value of 154° derived from the formula

---

*We note that the parasitic-modulation effect associated with the change in amplification may be reduced on increasing the internal decoupling of the TW QPA, i.e., increasing the attenuation of the backward-wave absorber in the delay system of the TW QPA. In order to weaken the effect of noise interference an external valve or circulator must be introduced close to the switch.

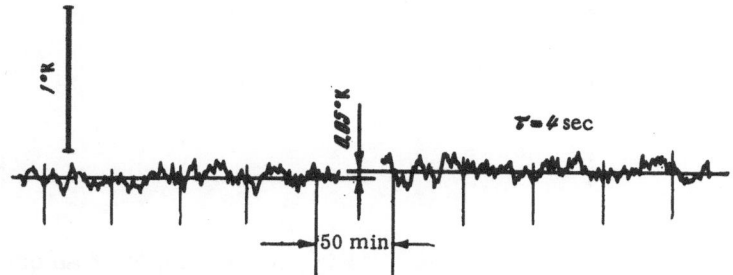

Fig. 9. Comparison of two parts of a "null" record separated by 50 min in time. Vertical lines are 1-sec time markers.

$$T_\mathrm{n} = T_\mathrm{a}\left(1 - \frac{1}{\alpha_\mathrm{dc}}\right) + \frac{T_0}{\alpha_\mathrm{dc}} + T_0\,\frac{1 - \eta_\mathrm{n}}{\eta_\mathrm{n}} + T_\mathrm{rec}\frac{1}{\eta}. \tag{9}$$

Here, η is the full efficiency of the tract (~1.1 dB); η_n is the efficiency of the noisy part of the tract (~1.0 dB); T_0 is the temperature of the space into which the undesirable waves are radiated (~30°K); T_a is the temperature of the antenna (~25°K); α_dc is the transitional attenuation of the directional couplers (~30 dB); and T_rec is the noise temperature of the amplifying tract, including the noise of the QPA and the subsequent mixing receiver (~30°K) (the value of T_n is reduced until the section after the screen).

As a result of the symmetry in the antenna-equivalent system, this radiometer has qualities which are very valuable for the type of observations being made. Figure 8 shows the value of $T_\mathrm{a} - T_\mathrm{e}$ as a function of the angle of elevation of the observed point. For comparison, this figure also shows the variations in background temperature obtained from measurements at the nearby wavelength of 5.3 cm [13]. We see from the figure that the effect of the background radiation is weakened by several orders of magnitude, facilitating measurements down to the lowest angles, at which the background radiation is extremely strong.

The radiometer output has a very stable "zero" over a long period. Figure 9 compares two parts of a record differing by 50 min in time. The "zero" drift in this time is no greater than 0.05°. Even unfavorable meteorological conditions (severe cloud cover, showers, etc.) caused no serious displacement of the "zero." This latter fact may be presumably explained by the circumstance that at a distance of several kilometers (typical distance to a cloud layer) the linear distance between the diagrams is measured in tens of meters. Clearly, at such a distance the radiation of the clouds and water vapor should still be quite strongly correlated with respect to intensity. The high sensitivity of the radiometer enabled sources producing an antenna temperature of the order of 0.1°K to be reliably recorded on a single record. For an effective antenna area of about 190 m² this corresponds to a flux of radio radiation equal to $1.5 \cdot 10^{-26}$ W/(m² · Hz). Figure 10 shows an example of an original record obtained from a relatively weak source of radio radiation, STA-102. The antenna temperature of this source is 0.12°K.

For an observation at culmination, the sources pass alternately through both positions of the polar diagram of the antenna. Figure 11 illustrates this for the case of the ZS273 source. The equivalence of the two diagrams may be used to increase the accuracy of observation of weak sources by a factor of two. By bringing the radio telescope to bear on the source first with one and then with the other possible mode of reception, we obtain a signal twice as great as that given by the ordinary "on — off" method. The radio-radiation fluxes of a number of quasistellar sources were determined at the wavelength of the radiometer by measurements made in February to June, 1966. In some of these systematic flux variations were observed. The scientific results of these observations were described in [14].

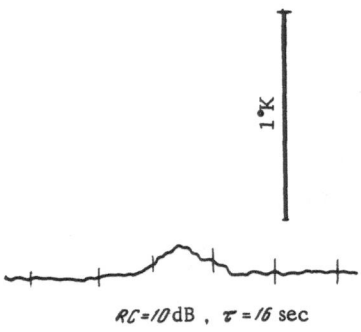

Fig. 10. Example of an original record for the passage of the STA-102 source of radio radiation.

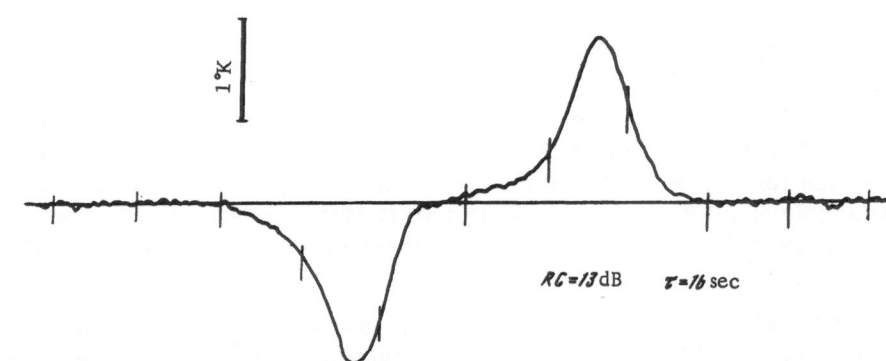

Fig. 11. Passage of the ZS273 source through both diagrams on observing at culmination.

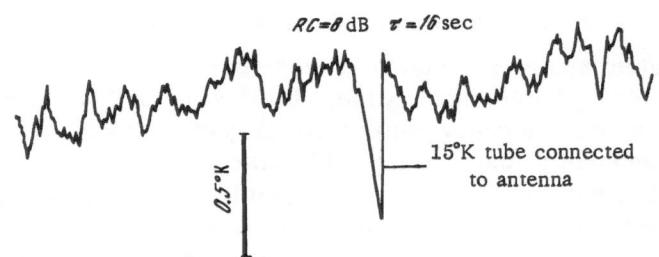

Fig. 12. Influence of the continuous spectrum on the output of the spectral channel.

The major project of the radiometer was the observation of the lines of excited hydrogen $n_{105} \rightarrow n_{104}$ (5762, 89 MHz). In the spectral channel a sensitivity of $\delta T = 0.04°$ was realized with a time constant of $\tau = 50$ sec, usually used for the observations. The zero line of the spectral channel was more stable than in the ΔF channel line. The influence of the continuous spectrum on the output of the spectral channel was weakened by at least 100 times. Figure 12 shows an oscillogram of the output of the spectral channel on applying a temperature step of 15°K to the input of the radiometer. Only at the initiation of the step, owing to the finite time constant of the control system, do we notice a surge of about 0.5°K. After developing a system with an accuracy of a few hundredths of a degree no displacement of the zero line was observed. The only fault in the radiometer was the slope of the zero line in the spectrograms, and no solution

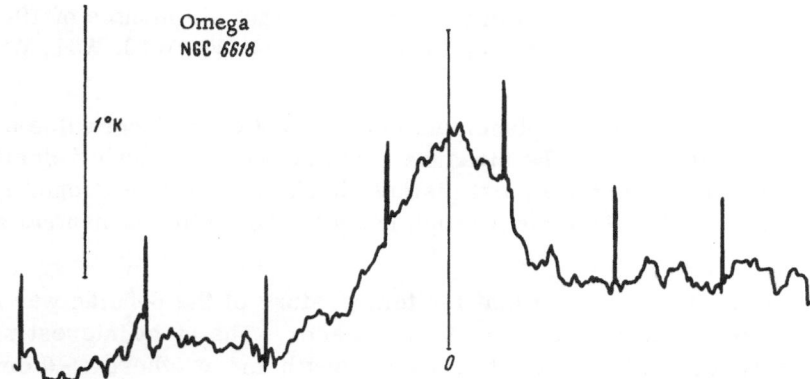

Fig. 13. Spectrogram of an original record of a line of the Omega nebula.
Vertical lines constitute frequency markers at 1 MHz intervals. The
large 0 mark is the calculated line frequency.

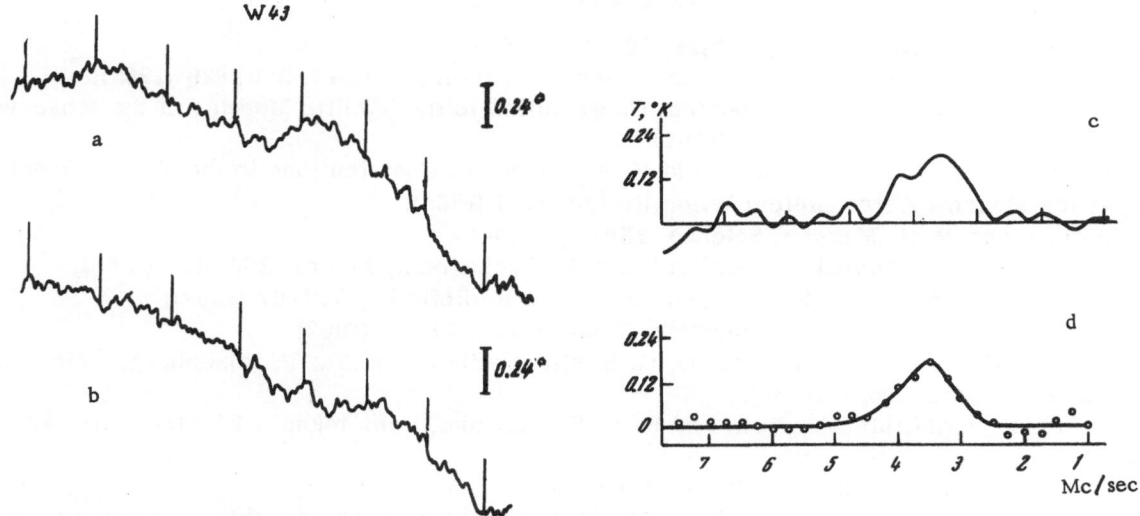

Fig. 14. Separation of a spectral line of the W43 source. a) Original spectrogram on receiving
the source radiation through an irradiator with a positive deviation from the line signal. Vertical lines constitute frequency markers at 1 MHz intervals; b) the spectrogram obtained on receiving the source radiation through an irradiator with a negative deviation from the line signal;
c) difference spectrogram of records a and b; d) average of nine pairs of records.

to this problem was ever found. Possible reasons for this phenomenon are set out in Section 4.
This slope was reduced to the comparatively small value of 0.04-0.05 deg/MHz as a result of
the measures taken. Figure 13 shows a spectrogram of the original recording of the $n_{105} \to n_{104}$
line in the Omega nebula. The slope of the zero line equals 0.04 deg/MHz. In observing weak
sources the observation procedure based on recording spectrograms by alternately tracking the
source first with one possible mode of the receiver and then with the other proved, as expected,
extremely effective. The difference spectrograms thus obtained (by pair analysis) were practically free from apparatus effects, while the spectral line under examination had twice the
value. Figure 14 illustrates the procedure for separating the line of the comparatively weak

W43 source.* By this method, spectral lines were reliably separated up to $T_l = 0.06°K$ (antenna temperature). As a result of observations made in the first six months of 1966 the $n_{105} \rightarrow n_{104}$ line was measured and its parameters determined [16] for the W3, W10, W31, W37, W38, W43, W49, and W51 emission nebulas.

Measurements of the Doppler displacements of the excited-hydrogen line enabled the velocities.of the nebulas to be calculated. The measurements showed that these velocities closely coincided with the radial velocities of the most distant clouds of neutral hydrogen lying on the path of the source, which indicated a close correlation between the motion of neutral and ionized hydrogen.

An extremely interesting fact was that the temperature of the nebulas was far lower than had been indicated by observations based on other methods. The most interesting conclusion from the study of spectral radiation both at this wavelength and at others (3-6 cm) is the absence of the theoretically predicted Stark broadening of the lines.

The authors wish to express their sincere thanks to A. E. Salomonovich for his great interest and attention regarding the present investigation, and also V. V. Zotov and L. M. Nagornykh for their practical help.

LITERATURE CITED

1. H. J. Ewen and E. M. Purcell, Nature, 168:356 (1951).
2. S. Weinreib, S. H. Barrett, M. L. Meeks, and J. C. Henry, Nature, 200:829 (1963).
3. R. L. Sorochenko and E. V. Borodzich, Contribution to the Twelfth Meeting of the Moscow Astronomical Union [in Russian] (1964).
4. Z. V. Dravskikh, A. F. Dravskikh, and V. A. Kolbasov, Contributions to the Twelfth Meeting of the Moscow Astronomical Union [in Russian] (1964).
5. B. Hoglund and P. G. Mezger, Science, 150:339 (1965).
6. H. E. Lilley, D. H. Menzel, H. Penfield, and B. Zuckerman, Nature, 209:468 (1966).
7. E. V. Borodzich and R. L. Sorochenko, Izv. VUZ, Radiofizika, 6:1167 (1963).
8. P. D. Kalachev and A. E. Salamonovich, Trudy FIAN, 17:13 (1962).
9. V. B. Shteinshleiger, O. A. Afanas'ev, G. S. Misezhnikov, and Ya. P. Rozenberg, Pribory i Tekh. Eksperim., No. 5 (1964).
10. N. V. Karlov, R. M. Martirosyan, and R. L. Sorochenko, Radiotekhn. i Elektron., 10:40 (1965).
11. V. S. Troitskii, Zh. Tekh. Fiz., 25:1426 (1955).
12. V. P. Bibinova, E. V. Borodzich, R. L. Sorochenko, and I. V. Shavlovskii, this volume, page 119.
13. R. W. De Grasse, D. C. Hogg, E. A. Ahm, and H. E. D. Scovil, J. Appl. Phys., 30:3013 (1959).
14. L. I. Matveenko, V. M. Kostenko, R. L. Sorochenko, and V. V. Zotov, Izv. VUZ, Radiofizika, 11:682 (1968).
15. G. Westerhout, BAN, 14:215 (1958).
16. V. M. Gudnov and R. L. Sorochenko, Astron. Zh., 44:1001 (1967).

---

*Number based on the Westerhout catalog [15].

POSSIBILITY OF USING THE COMPENSATION METHOD FOR SPECTRAL MEASUREMENTS

É. V. Borodzich, Yu. S. Rusinov, and R. L. Sorochenko

Since the construction of radio receiving apparatus with a noise temperature of the order of 5-10°K, based on the maser amplifier, is now a real possibility, the question of eliminating the switch with its 0.3-0.5 dB (20-30°K) loss has acquired fundamental importance. If it is taken into account that the theoretical sensitivity of a compensating radiometer exceeds that of a modulation radiometer by a factor of $\pi/\sqrt{2}$ (2.23), the development of switchless radio-astronomical receivers becomes definitely feasible.

As is well known, the main shortcoming of compensating systems is the difficulty of realizing their high theoretical sensitivity, due to instability of the transmission coefficient of the amplifying line. For spectral measurements, however, one can diminish appreciably the effect of instabilities in the broadband part of the radiometer by using certain circuits. We shall consider the radiometer block diagram shown in Fig. 1.

The operation of the radiometer is based on the joint use of a quasi-null method, which enables one to maintain the sum of external and internal radiometer noises in a broad band constant with time, and a differential method of distinguishing the spectral signal by comparing signals in the broad and narrow bands. The high-frequency signal contained in a broad band ΔF is amplified, detected by the detector D_0, and then used for control by the compensation-noise generator (NG). The signal of this generator is supplied to the radiometer input by an auxiliary radiator; in this case, it passes through the same components as the signal under investigation.

The feedback loop so formed serves to maintain a constant output noise signal:

$$P_{\text{out}} = k\Delta F \overline{T(f,t)}\,\overline{N(f,t)} = \text{const},\tag{1}$$

where $\overline{T(f,t)}$, $\overline{N(f,t)}$ are the averaged noise temperature and gain factor in the band ΔF; k is Boltzmann's constant.

At the same time, m subbands, having the band widths Δf_i, are isolated from the band ΔF by using the narrow-band filters Fi_i. The noise signals contained in these subbands are each detected by the detectors D_i and then applied to the balancing meshes B_i, which are also connected to D_0.

In the absence of the spectral signal under investigation, the initial balance of the analyzers reaches the equality

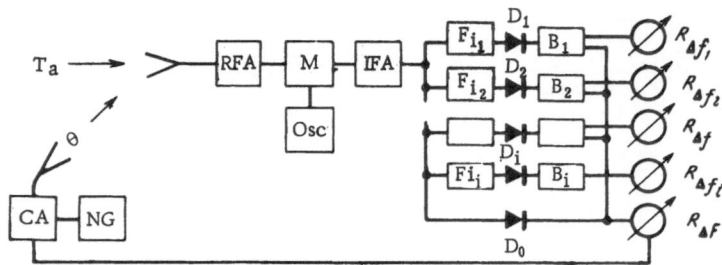

Fig. 1. Radiometer block diagram. $R_{\Delta f_i}$ are channel recorders; Fi_i are narrow-band filters; B_i are balancing meshes; CA is controlled attenuator; NG is a noise generator; Osc is a fine-tuned beat oscillator; RFA = radio frequency amplifier; M = mixer; IFA = intermediate frequency amplifier.

$$k\Delta F \overline{N(f, t)}\ \overline{T(f, t)} = \alpha_i k\Delta f_i N(f_i, t)\, T(f_i, t), \tag{2}$$

where $\alpha_i \approx \underline{\Delta F / \Delta f_i}$, as a result of which the output of each spectral channel reads "zero." In Eq. (2), $N(f_i, t)$ is the gain of the channel with the band Δf_i at the frequency f_i, and $\overline{T(f_i, t)}$ is the noise temperature of the radiometer input at this frequency, determined by the equation

$$T(f, t) = T_a(f, t) + \theta(t) + \frac{1 - L}{L} T_0 + \frac{T_m(f, t)}{L} + \frac{T_s(f, t)}{L\, n(f, t)}. \tag{3}$$

In this equation, $T_a(f, t)$ is the noise temperature due to the effect of radiation, including the signal under investigation; $\theta(t)$ is the noise temperature of the compensating noise, which has a continuous spectrum; $T_m(f, t)$ and $T_s(f, t)$ are the noise temperatures of the maser amplifier and superheterodyne receiver, respectively; $n(f, t)$ is the gain of the maser amplifier; and L is the transmission coefficient of the high-frequency line.

Since $T_m(f, t)$ is determined by the spin temperature and ohmic loss in the input components of the maser amplifier has a very weak frequency gradient, the internal noise of the radiometer can be assumed to be independent of frequency if the gain $n(f, t)$ is sufficiently high.

In this case, owing to the operation of the feedback loop and balancing mesh, changes in $\overline{T(f, t)}$ and $\overline{N(f, t)}$ do not upset equality (2) and, hence, do not cause zero drift in the different channels. At the same time, when a signal under investigation having a nonuniform spectral density arrives at the radiometer input, the corresponding channels record an excess in the spectral density of noise at the channel frequency relative to the average level. A spurious signal, i.e., zero drift in one or more channels in the case where the noise under investigation is spectrally uniform, will occur when the pass band is deformed with time.

It can be shown that the amplitude of the spurious signal in the i-th channel for the time of measurement δt is proportional to the product of the total noise temperature and the difference between the derivatives, with respect to time, of the gain factors in the broad and i-th bands:

$$\delta T_{s.d.} = \frac{\overline{T(f, t)}}{\overline{N(f, t)}} \left(\frac{d\overline{N(f, t)}}{dt} - \frac{dN(f_i, t)}{dt} \right) \delta t. \tag{4}$$

The difference, enclosed by parentheses in Eq. (4), should amount to less than $\frac{1}{10}$ of the change in each of the gain factors $\overline{N(f, t)}$ and $N(f_i, t)$ in particular, which correspondingly decreases the spurious signal in this method of reception as compared with the usual compensation method.

It should be noted that in the quasi-null spectrometer, the feedback loop and the balancing mesh equally promote the maintenance of equalities (2) and (3). The effect of inaccuracy in adjusting the feedback loop is decreased by the differential method of isolating the spectral signal and vice versa. Such a circuit provides substantially more reliable protection against changes in the intensity of the continuous spectrum than exists in a constant-current radiometer, in which only the differential method is used [1]. This is especially important at wavelengths of 3 cm or less, where the external noises exceed the internal ones [2].

A fundamental property of the quasi-null spectrometer is the maintenance of a constant noise level in the high-frequency line. This diminishes substantially the effect of interference phenomena on the operation of the spectrometer. As is well known, such phenomena occur when the high-frequency line, connecting the antenna to the amplifier input, is detuned. Interference of noises radiated by the amplifier input toward the antenna and reflected back, as well as noises received by the antenna and reflected from the amplifier input, results in variation of the spectral density of noise with frequency at the radiometer input.

According to [3], this variation may be determined by the equation

$$W\left(f\right) = 4k\rho \left\{ \frac{k_{m}T_{1}\left[1 + (k_{a}^{2} - 1)\sin^{2}\frac{2\pi fl}{v_{p}}\right] + T_{a}k_{a}}{(k_{m} + k_{a})^{2} + (1 - k_{m}^{2})(1 - k_{a}^{2})\sin^{2}\frac{2\pi fl}{v_{p}}} \right\},$$ (5)

where $T_1 \approx T_m$ is the noise temperature radiated by the amplifier input in the direction of the antenna; k_m and k_a are, respectively, the standing-wave coefficients of the amplifier input and antenna; ρ and l are the wave impedance and length of the line, and v_p is the phase velocity of propagation in the line.

The periodic component of this equation characterizes the spurious signal and on the temperature scale has the form:

$$T\left(f\right) \approx 4T_{m}\frac{(k_{a}^{2} - 1)k_{m}}{(k_{a} + k_{m})^{2}}\sin^{2}\frac{2\pi fl}{v_{p}} - 4T_{a}\frac{k_{a}(1 - k_{m}^{2})(1 - k_{a}^{2})\sin^{2}\frac{2\pi fl}{v_{p}}}{(k_{a} + k_{m})^{4}}.$$ (6)

The elimination or reduction to reasonable limits of this effect is one of the most serious problems of the technique of spectral measurements. The difficulties arising in this case, as well as certain practical results obtained in modulation-type spectrometers, are described in [4, 5].

In Eq. (6), the second term, which characterizes the spurious signal due to change in intensity of the external radiation ΔT_a, is highly undesirable. Since this radiation is the subject of the investigations, eliminating or allowing for the quantity

$$\Delta T_{a}\left(f\right) \approx \Delta T_{a}\frac{4k_{a}(1 - k_{m}^{2})(1 - k_{a}^{2})\sin^{2}\frac{2\pi fl}{v_{p}}}{(k_{a} + k_{m})^{4}}$$ (7)

is a necessary condition for the reception of weak spectral signals. However, this condition is met in the quasi-null radiometer, where, owing to the feedback loop, all changes in external radiation are compensated by locally generated noise. As a result, the spurious signal from interference of noises is decreased to the value

$$\delta T_{s.i.} \approx T\left(f\right)\frac{d}{dt}\overline{\frac{N\left(f, t\right)}{\widetilde{N}}}\,\delta t.$$ (8)

In this case, the undesirable spectral dependences determined by Eq. (6) can be eliminated by initial balancing of the channels according to (2). Maintenance of this state is ensured provided that the change in the gain factor, as well as deformation of the radiometer characteristic, is a slower process than the taking of the spectrogram. However, such a condition is not excessive in view of the multichannel reception and substantial shortening of the required exposure, due to elimination of the switch.

LITERATURE CITED

1. W. E. Selove, Rev. Sci. Instr., 17:268 (1946).
2. M. E. Tiuri, IEEE Mil., 8:264 (1964).
3. V. S. Troitskii, Zh. Tekh. Fiz., 25:1426 (1955).
4. V. M. Gudnov, I. M. Goryachev, V. A. Kolbasov, G. S. Misezhnikov, R. L. Sorochenko, B. V. Sestroretskii, and V. B. Shteinshleiger, This volume, p. 1.
5. V. P. Bibinova, E. V. Borodzich, R. L. Sorochenko, and I. V. Shavlovskii, This volume, p. 119.

AN 8-cm RADIOMETER WITH A QUANTUM PARAMAGNETIC AMPLIFIER

L. I. Matveenko, G. S. Misezhnikov, M. M. Mukhina, and V. B. Shteinshleiger

Introduction

In recent years radio-astronomical methods of research have become widespread in astrophysics. These methods make it possible to record from cosmic objects radiation in a wide range of radio frequencies from which conclusions about their physical properties are reached. In the majority of cases, the intrinsic radiation of the radio sources is not great, and this imposes exacting demands on the sensitivity of radio telescopes. The minimum detectable flux density is related to the parameters of the radio telescope as follows:

$$\delta S \sim \frac{\delta T}{A_{\text{eff}}},\qquad(1)$$

where δT is the fluctuational sensitivity of the radiometer, and A_{eff} is the effective area of the antenna.

As we know, the higher the fluctuational sensitivity of a radiometer the lower the temperature of the noises at its input. As a result of the successes of quantum radiophysics low-noise quantum paramagnetic amplifiers (QPA's) have been developed in recent years. In particular, traveling-wave QPA's of similar design for the 5- and 8-cm bands have been developed [1,2], but so far they have not been used in radio-astronomical measurements in the USSR. In 1962 a start was made on developing a radiometer with an 8-cm range, running-wave quantum paramagnetic amplifier. The present article is devoted to the description of this radiometer and the first results of observations.

1. Choice of Design for the Radiometer

The fluctuational sensitivity of a radiometer due to statistical fluctuations is defined by the expression

$$\delta T = \alpha \frac{T_{\text{n}}}{\sqrt{\Delta f \tau}},\qquad(2)$$

where T_{n} is the temperature of the noises entering the input of the radiometer; Δf is the transmission band of the radiometer to the square-law detector; τ is the integrating chain time constant; and α is a coefficient defined by the type of radiometer. In radio astronomy the following main types of radiometers have come into use for measurements in the continuous spectrum: compensation, correlation, and modulation.

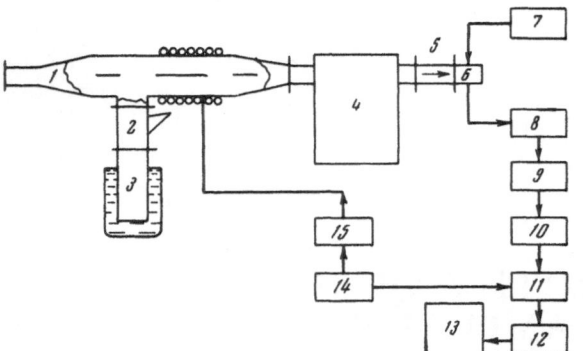

Fig. 1. Block diagram of a radiometer with a QPA. 1) Amplitude modulator; 2) smooth attenuator; 3) load equivalent; 4) QPA; 5) isolator; 6) mixer; 7) heterodyne; 8) intermediate-frequency preamplifier; 9) intermediate-frequency amplifier and detector; 10) low-frequency amplifier; 11) synchronous detector; 12) dc amplifier; 13) recording apparatus; 14) modulation frequency generator; 15) power amplifier.

The greatest fluctuational sensitivity is found in the compensation method, for which $\alpha = 1/\sqrt{2}$. The minimum belongs to the modulation method; depending on the form of modulation and demodulation, α lies within the limits $\pi^2/8 - 2$. In the case of the correlation type $\alpha = 1$ [3]. However, despite the high degree of fluctuational sensitivity, the compensation method has a number of defects, namely, that in order to realize its sensitivity it is necessary to ensure the stability of the amplification coefficient $\Delta k/k$, corresponding to

$$\frac{\Delta k}{k} < \frac{\delta T}{T_n} \approx 10^{-4}, \quad \text{and since } \delta T \ll T_n, \text{ its amp-}$$

litude characteristic must be linear in a broad dynamic range. In the correlation method the signal recorded is proportional to the correlation part of two independently received signals, which significantly weakens the effects of instability in the amplification coefficient. However, even in this case, the requirement of linearity in the amplitude coefficient is kept, and moreover the condition of large decoupling between the inputs of the radiometer is imposed as well. The modulation method is free from these defects, but makes severe demands on the modulating apparatus. Ferrite switches and rectifiers, which have become widespread in recent years, make it possible to exclude "parasitic" effects quite simply, and to realize the fluctuational sensitivity of this method. Thus a radiometer with amplitude modulation, having a higher degree of technical sensitivity, was chosen.

2. Radiometer with a Quantum Paramagnetic Amplifier

The radiometer consists of the following main units: an amplitude modulator, a quantum paramagnetic amplifier, a transformation unit, a low-frequency part, and a supply. The block diagram of the radiometer is shown in Fig. 1. For greater reliability the design of the units has been simplified as much as possible. The theoretical schemes and the requirements for the units are considered below.

Input Circuits of the Radiometer. In order to increase the fluctuational sensitivity of the radiometer, attention must be given to reducing its noise temperature. In accordance with the block diagram shown in Fig. 1, the noise temperature of the radiometer is defined by the following expression (it is assumed here that the high-frequency channel is sufficiently well matched and that the effect of mismatching on the noise temperature can be neglected):

$$T_n = m\left\{T_a + T_e + \frac{\beta_{m.e} + \beta_{m.a}}{\beta_{m.e}\,\beta_{m.a}}\left[\frac{T_m}{G\beta} + \frac{T_{QPA}}{\beta} + (1-\beta)\,T_0\right] + T_0(2 - \beta_{m.e} - \beta_{m.a})\right\} + (0.5 - m)\,T',$$

(3)

where T_a, T_e, and T_0 are the noise temperatures of the antenna, the equivalent, and the surrounding medium, respectively; T_m and T_{QPA} are the noise temperatures of the mixer and the quantum paramagnetic amplifier; $\beta_{m.a}$ and $\beta_{m.e}$ are the direct transmission coefficients of the modulator at the input of the antenna and of the equivalent, respectively; β is the transmission coefficient of the high-frequency tract between the modulator and the quantum paramagnetic amplifier; G is the amplification coefficient of the QPA with respect to the power; m is a coefficient defined by the relative time for switching on the antenna and the equivalent; and T' is the noise temperature of the radiometer at the moments of switching.

The noise temperature of the antenna T_a, determined for the zenith, is defined by the irradiation on it and by the temperature of the cosmic radio-radiation background. The temperature of the equivalent, T_e should, according to [4], be made equal to the temperature of the antenna in order to reduce the effect of instability in the radiometer coefficient. Thus the noise temperature of the radiometer can be reduced by lowering the noise temperature of the mixer and the losses in the high-frequency tract and applying a low-noise amplifier with a high amplification coefficient.

Further, the modulator must have a low switching time. In the case of a low temperature T', at switching the radiometer can be disconnected by means of disabling pulses. These requirements were taken into account in designing the high-frequency part of the radiometer.

Modulator. The amplitude modulator is the most important element in the radiometer; on it depends the realization of the fluctuational sensitivity. Research on the operation of amplitude modulators constructed of crystal diodes or ferrite working in a longitudinal magnetic field has shown that the damping of the latter does not significantly exceed the damping of the former; but the modulator made of ferrite is more stable in its operation and does not require tuning, and its level of parasitic modulation is significantly lower and, in practice, does not depend on the ambient temperature. In view of this, the ferrite modulator was used in the radiometer.

Let us consider how the parasitic signal is connected with the parameters of the modulator. We represent the modulator in the form of a four-arm circulator connected in accordance with the scheme shown in Fig. 2. We denote by β_n the direct transmission between the nearest inputs, by γ the inverse attenuation, by T_n the noise temperatures at its inputs, and by $\Gamma_n = (k_n - 1)/(k_n + 1)$ the reflection coefficients. If we neglect the interference terms and the terms corresponding to double reflection, the signals at the output of the QPA will be equal to:

antenna switched on:

$$P_a \propto [T_a(1 - \Gamma_a^2)\beta_{a\,\mathrm{outp}} + T_e(1 - \Gamma_e^2)\gamma_{e\,\mathrm{outp}} + T_l(1 - \Gamma_l^2)\Gamma_a^2\beta_{l.a}\beta_{a.\mathrm{outp}} +$$
$$+ T_{\mathrm{QPA}}(1 + 2\Gamma_e\sqrt{\beta_{e\,\mathrm{outp}}\gamma_{e\,\mathrm{outp}}} + \Gamma_e^2\beta_{e\,\mathrm{outp}}\gamma_{e\,\mathrm{outp}}) + T_0(1 - \beta_{a.\mathrm{outp}})]G_a, \tag{4}$$

equivalent switched on:

$$P_e \propto [T_e(1 - \Gamma_e^2)\beta_{e\,\mathrm{outp}} + T_a(1 - \Gamma_a^2)\gamma_{a\,\mathrm{outp}} + T_l(1 - \Gamma_l^2)\Gamma_e^2\beta_{l e}\beta_{e\,\mathrm{outp}} +$$
$$+ T_{\mathrm{QPA}}(1 + 2\Gamma_a\sqrt{\beta_{a\,\mathrm{outp}}\gamma_{a\,\mathrm{outp}}} + \Gamma_a^2\beta_{a\,\mathrm{outp}}\gamma_{a\,\mathrm{outp}}) + T_0(1 - \beta_{e\,\mathrm{outp}})]G_d. \tag{5}$$

Obviously the signal at the output of the radiometer will in the case of square-law detecting be proportional to the difference between these quantities, i.e.,

$$P = T_a(1 - \Gamma_a^2)\beta_{a\,\mathrm{outp}} - T_e(1 - \Gamma_e^2)\beta_{e\,\mathrm{outp}}\frac{G_e}{G_a} + T_e(1 - \Gamma_e^2)\gamma_{e\,\mathrm{outp}} -$$
$$- T_a(1 - \Gamma_a^2)\gamma_{a\,\mathrm{outp}}\frac{G_e}{G_a} + T_l(1 - \Gamma_l^2)\left(\Gamma_a^2\beta_{l.a}\beta_{a\,\mathrm{outp}} - \Gamma_e^2\beta_{l e}\beta_{e\,\mathrm{outp}}\frac{G_e}{G_a}\right) +$$
$$+ T_{\mathrm{QPA}}\left[(1 + 2\Gamma_e\sqrt{\beta_{e\,\mathrm{outp}}\gamma_{e\,\mathrm{outp}}} + \Gamma_e^2\beta_{e\,\mathrm{outp}}\gamma_{e\,\mathrm{outp}}) -\right.$$
$$\left. - (1 + 2\Gamma_a\sqrt{\beta_{a\,\mathrm{outp}}\gamma_{a\,\mathrm{outp}}} + \Gamma_a^2\beta_{a\,\mathrm{outp}}\gamma_{a\,\mathrm{outp}})\frac{G_e}{G_a}\right] + T_0\left[(1 - \beta_{a\,\mathrm{outp}}) - (1 - \beta_{e\,\mathrm{inp}})\frac{G_e}{G_a}\right]. \tag{6}$$

As can be seen from this expression, when $T_a = T_e$ the signal at the output of the radiometer is nonzero. This parasitic signal is caused in the general case by the following factors: (a) mismatching of the inputs of the modulator; (b) a difference between the parameters of the modulator in the different arms; and (c) dependence of the amplification factor of the QPA on the matching of its input.

Fig. 2. Amplitude modulator (external view, equivalent scheme, and form of modulation).

The parasitic signal can be compensated for by an appropriate change in the temperature of the equivalent but, since its magnitude varies with variation in the parameters of the modulator, the high-frequency channel, and the amplification factor, the balance will be unstable. We shall assume that the antenna and the equivalent are matched, the modulator is symmetric and exhibits only slight losses, and $T_a = T_e$. In this case it follows from (6) that the parasitic signal will be given by a quantity roughly equal to $(T_a + T_{QPA}) \Delta G / G$. Since $T_a + T_{QPA} \approx 100°K$, the parasitic signal can clearly attain a considerable magnitude. We used a traveling wave QPA working in an absorption mode in accordance with a pumping signal and having a high degree of decoupling in the reverse direction. It can easily be shown that in this case

$$\Delta G/G \approx 2\,(\Gamma_a - \Gamma_e) \sqrt{GL\gamma}\, \Gamma_{outp\,QPA},$$

where L and γ are the damping of the signal in the reverse direction in the QPA and the modulator, respectively. When L = −40 dB, γ = −25 dB, $\Gamma_{outp\,QPA}$ = 0.2 and the standing-wave ratio of the outputs of the modulator is equal to about 1.2, the effect of ΔG can be neglected.

In this case after some straightforward transformations the parasitic signal can be represented in the form

$$P \approx (T_a - T_0)(\beta_{a\,outp} - \beta_{e\,outp}) + T_a\,(\gamma_{a\,outp} - \gamma_{a\,outp}) + T_e\,(\Gamma_a^2 - \Gamma_e^2) +$$
$$+ T_{QPA}\,[(\Gamma_e^2 \gamma_{e\,outp} - \Gamma_a^2 \gamma_{a\,outp}) + 2\,(\Gamma_e \sqrt{\gamma_{e\,outp}} - \Gamma_a \sqrt{\gamma_{a\,outp}})].$$

It follows from this expression that given sufficiently good decoupling of the modulator (around 20 dB) the effect of the second and fourth terms can be neglected. In that case the principal parasitic signal will be given by the third term in the equation, since $T_n = 300°K$. In order to reduce its effect it is obviously necessary to match carefully the inputs of the antenna and the equivalent and as far as possible to make them identical. In order that the contribution of this term should not exceed 0.5°K the matching of the inputs should be of the order of s.w.r. $\approx$ 1.2, and they should differ from one another by not more than a few percent. For it to be possible to neglect the effect of the first term, the difference in the damping should not exceed a few percent when $\beta_n = 0.96$. The conditions we have considered were taken into account in designing the modulator.

The general appearance of the modulator is shown in Fig. 2. The ferrite is located in the circular waveguide, in the magnetic field of the solenoid. Variation in the magnetic field with the frequency of modulation leads, on account of the Faraday effect, to rotation of the plane of polarization of the signal through ±45°. The length of the ferrite was chosen in such a way

Fig. 3. Traveling wave quantum paramagnetic amplifier.

that the rotation attained saturation at an angle of 45° and thus created modulation which was close to rectangular (shown in Fig. 2) given a sinusoidal modulation voltage. Thus the sensitivity is increased and we exclude the effect of instability in the amplitude of the modulating voltage. By tuning the modulator, a decoupling of not less than 25 dB was achieved between its inputs along with matching of s.w.r. = 1.02 with respect to the antenna input and s.w.r. = 1.045 with respect to the input of the equivalent in the 20-MHz radiometer transmission band. In order to reduce losses, the ferrite was dried and placed in a hermetically sealed polystyrene container. The losses in the modulator through the antenna input were distributed in the following way: waveguides, 0.06 dB; absorbent plate, 0.01 dB; and ferrite, 0.08 dB. Total losses equaled 0.15 dB. Damping at the input of the equivalent did not exceed 0.18 dB. The measured noise temperature of the modulator was equal to 10-15°K. The power of the modulating voltage amounts to 6.5 W.

Load Equivalent. The temperature of the equivalent must be equal to the noise temperature of the antenna. In our case the noise temperature of the antenna at medium angles of elevation is about 80°K. The standard equivalent in the form of a small horn antenna is not very acceptable in this case because of its low directivity, which leads to reception of interference and a reduction in sensitivity. In view of this the equivalent was made in the form of a waveguide of cross section 10×60 mm loaded with a wedge of polyiron. In order to reduce heat conductivity and damping, the waveguide was made of silvered metallized micarta. The flange of the load is grounded and in order to ensure that it is hermetically sealed it is joined to the modulator by a thin Teflon gasket. Measurement of the absorption of the waveguide by the short-circuit method showed that it does not exceed 0.5%. The load was placed in liquid nitrogen, the boiling point of which is 78°K. In order to exclude variations in the load impedance due to the liquefaction of the air inside it, the load was filled with gaseous helium coming out of a small rubber ball placed in an opening in the broad wall of the waveguide. The s.w.r. of the load within the frequency range of the radiometer is no worse than 1.05. Thus the noise temperature of the load at its output flange did not exceed 80°K. The temperature of the equivalent was regulated by means of a smooth attenuator with a low damping, 0-0.5 dB. The matching of the whole system gives a s.w.r. ≤ 1.15.

The noise temperatures of the radiometer and the calibration step were measured by means of a load similar to the equivalent. The load temperature was changed by placing it in liquid nitrogen, melting ice, and boiling water. The s.w.r. of the load in the radiometer band is less than 1.05 and is practically constant with respect to temperature changes.

Quantum Paramagnetic Amplifier. In order to reduce the noise temperature in the radiometer, a traveling-wave quantum paramagnetic amplifier is used; its external appearance is shown in Fig. 3. The amplifier is installed in a horizontal position and is joined to the modulator by a rotating junction. The QPA, in comparison with an amplifier of the resonator type, has a wider transmission band (~20 MHz), a lower level of intrinsic noise because of the absence of a circulator at the input, and a more stable amplification factor. The active material

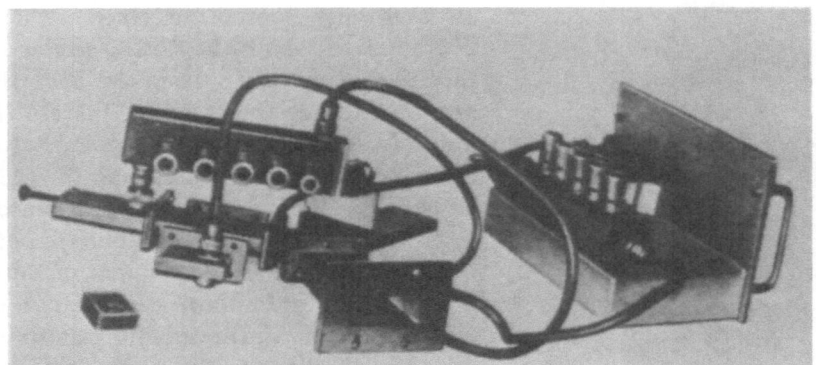

Fig. 4. General view of the frequency transformation block.

is ruby. The angle between the direction of the trigonal axis of the crystal and the direction of the external magnetic field is 90°. The frequency of the signal corresponds to the transition between the first and second energy levels, and the angle of pumping to that between the first and fourth. This mode of working makes it possible to obtain the maximum amplification factor [1].

Research on the dependence of the amplification factor on the concentration of chromium in the ruby has shown that in this mode of operation the greatest amplification at 8 cm is attained at a concentration of Cr^{3+} equal to 0.036%; the coefficient of inversion in this case is equal to 4.5. The ruby crystals are placed on both sides of the pin-retarding system, the length of which is 165 mm. In the transmission band of the system (~ 150 MHz) the retardation of the wave in relation to the group velocity is equal to 100, and the intrinsic losses amount to 5 dB. At a temperature of 4.2°K, the matching of the input and output is equal to s.w.r. ≈ 1.5. In order to absorb the reverse wave, use is made of ferrite rectifier elements in the form of plates located under ruby rods in both sides of the retarding system. The best rectifying properties at helium temperatures in the 8-cm range have been shown by measurements to belong to polycrystalline iron — yttrium ferrite with garnet structure. A rectifying system with co-axial supply cables is placed in a metal cryostat between the poles of an external permanent magnet. The cryostat enables the QPA to work for 8 hours.

Putting a ruby rod on both sides of the retarding system has made it possible to amplify the QPA to 40% above the level reached using one side only and to obtain, when T = 4.2°K, an amplification of 32 dB with losses in the retarding system for the direct wave of 12 dB. The total amplification of the QPA is 20 dB in the band around 20 MHz. The reverse damping of the QPA is 40 dB. The intrinsic noise temperature of the QPA does not exceed 15°K. If required, the amplification factor can be increased to 35 dB in the band around 15 MHz by reducing the boiling point of the liquid helium to 2°K by removing its vapors. The amplifier allows tuning of ± 50 MHz by appropriate changes in the magnetic field intensity, without changing the frequency of pumping.

The transformer block (Fig. 4) consists of the mixer, heterodyne, and the preamplifier and principal amplifier of the intermediate frequency. At the output of the latter there is a square-law detector. The block is mounted directly on the quantum paramagnetic amplifier; this reduces the additional losses at high frequencies and eliminates rotating or flexible connections both at high and intermediate frequencies. The intrinsic noises of the transformation block together with the mirror channel are equal to 1350°K. The frequency characteristic is close to rectangular and ensures a transmission band, with respect to the half-power level, of roughly 25 MHz. The mixer is single-cycle and works with a D405 diode. The signal of the heterodyne is fed into the mixer through a directed coupler.

Fig. 5. General view of the low-frequency part of the resonator.

At the input of the mixer, a circulator is mounted as a rectifier; it prevents the heterodyne signal from getting into the QPA.

The intermediate-frequency preamplifier is connected directly to the output plug of the mixer. The input stages of this amplifier are collected at 6S3P and 6S4P tubes switched on in a cascade circuit.

As square-law detector a 6Zh1P electron tube in a diode circuit is used. The linearity of the amplitude characteristic of the detector is preserved up to 0.1 V at its input. The linearity of the radiometer as a whole was checked against a matched load analogous to the load equivalent, the temperature of which was made equal to the temperature of liquid nitrogen, melting ice, and boiling water, and also against radio sources with known radio radiation beams.

Low-Frequency Part of Radiometer. The general appearance of the low-frequency part of the radiometer is shown in Fig. 5. It consists of a narrow-band low-frequency amplifier tuned to a modulation frequency F = 178 Hz. Experience has shown that extending the transmission band of a low-frequency amplifier in order to ensure complete transmission of the modulated signal does not lead to the expected gain in sensitivity both because of induced signals and because of the complications involved in ensuring that the harmonics of the signal and the reference voltage are cophased. So the transmission band of the amplifier was made sufficiently narrow: $\Delta F = 10$ Hz. Further narrowing of the band is inadvisable, since instability may appear here in the phase of the amplified voltage, especially when the amplification factor is changed by means of attenuators. In this case the requirements imposed on the stability of the modulation-frequency generator increase. The synchronous detector is mounted on a double 6Kh2P diode. Balancing it eliminates the effect of instability in the reference voltage. Thus a variation of 3 dB in the reference voltage ($U_{ampl} = 11$ V) goes practically undetected at the output of the radiometer.

The dc amplifier is mounted in accordance with the usual bridge circuit. The reference-voltage generator and the power amplifier generate the sinusoidal voltage of the modulation frequency F = 178 Hz. The modulation frequency is chosen with a view to eliminating induced signals from the supply system. The sinusoidal character of the voltage simplifies phasing and symmetrizing the voltage.

Supply Blocks. Except for the modulation-frequency generator block, the radiometer is supplied from specially developed anode voltage stabilizers with a stabilization coefficient of 2500, for a 10% variation in the voltage of the supply system. The external appearance of the stabilizer is shown in Fig. 5, and its circuit in Fig. 6. In order to reduce induced signals from the supply system a type V525 dc-filament voltage stabilizer is used in the radiometer. The modulation-frequency generator is supplied by the anode from a stabilizer with a stabilization coefficient equal to about 200.

Switching on the QPA significantly reduced the intrinsic noises of the radiometer. Thus, whereas the noise temperature of the radiometer without the QPA (without the mirror channel)

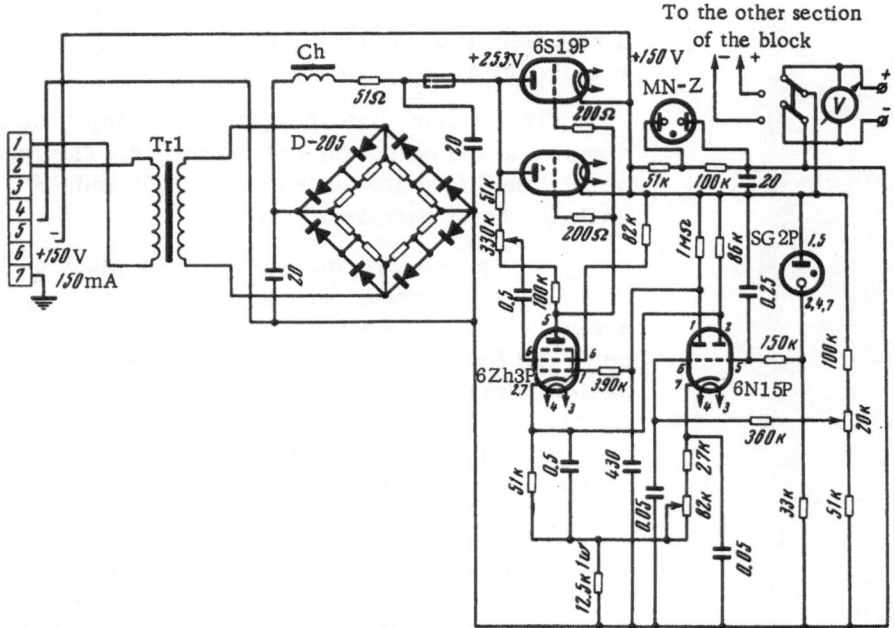

Fig. 6. The theoretical circuit of the anode voltage stabilizer.

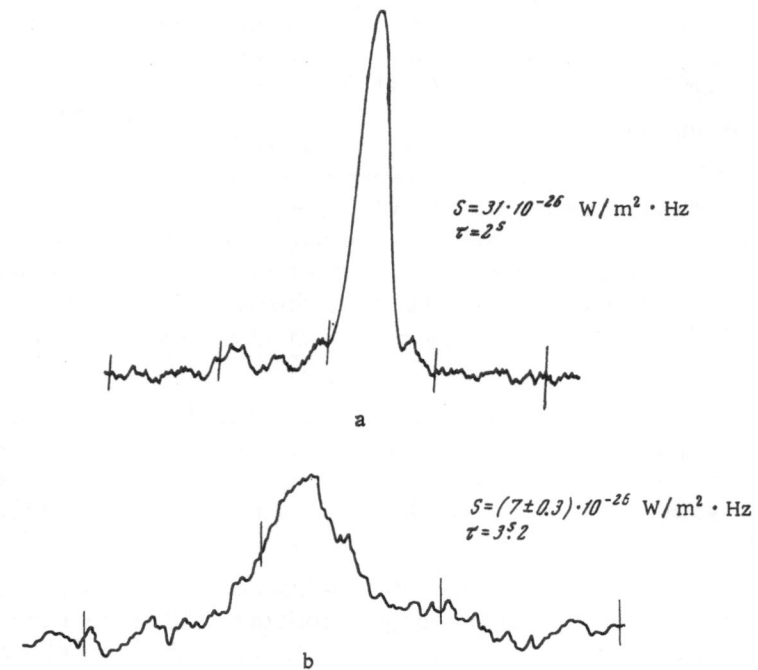

Fig. 7. Readings of the radio sources: a) 3S 273; b) 3S 84.

was 2700°, the radiometer with the QPA had a temperature of 50°K, i.e., less than one fiftieth of the value in the other case.

    As was to be expected, such a reduction in the noise temperature revealed effects (parasitic modulation, induced signals, etc.) which had previously been concealed in noise. Additional tuning of the high-frequency channel, s.w.r. < 1.15, filling of the waveguide of the load equivalent with gaseous helium, and supplying the filaments of the radiometer with very stable direct current made it possible to reduce these effects to the level of the noise path and reduce drift of the amplification factor to 2% per hour.

In the process of tuning the radiometer it was discovered that when the time constant τ varied from 0.5 to 2 sec the sensitivity of the radiometer with the QPA increased more rapidly (by about 30%) than could be expected from the relationship $\tau^{-\frac{1}{2}}$. This property is evidently connected with the high-frequency component of the fluctuations (> 1 Hz) caused by helium boiling in the cryostat.

In the radiometer a fluctuational sensitivity of $\delta T_p = 0.024°K$ was achieved by calculation. In this, the total noise temperature of the radio telescope, according to the measurements, was taken to be $T_n = 120°K$, the transmission band of the radiometer was $\Delta f = 20$ MHz, $\tau = 3.2$ sec, with the modulation being rectangular, the demodulation sinusoidal. The measured sensitivity in relation to sources with known radio-radiation flux densities, and by calibration from a standard load, was found to be equal to $\delta T = 0.024 \pm 0.003°K$.

3. Results of Observations

The use of the QPA has improved the fluctuational sensitivity of the radio telescope with respect to the flux by about ten times. Its sensitivity was close to the calculated value and was $\delta S = 1.5 \cdot 10^{-27}$ W/m^2 · Hz given the integrating [5] apparatus time constant $\tau = 3.2$ sec. By way of example, Fig. 7 shows the readings of the radio sources 3S 84 and 3S 273, for which the flux densities of their radio radiation [6] are 7 ± 0.3 and 31, respectively. In October 1963 radio astronomy observations were begun. Readings of the Crab nebula showed that its right ascension, calculated for the epoch 1950.0, was $\alpha = 5^h 31^m 29.7^s \pm 0.7^s$, i.e., displaced by $\Delta \alpha = -25 \pm 10''$ relative to the center of the double star to the west. The diameter of the nebula, calculated from the half-luminance level, assuming Gaussian distribution, was equal to $\varphi = 3.27 \pm 0.05'$. But the total extension of the region of radio emission in the direction of right ascension (at the 1% level) is equal to $6 \pm 1'$ and with accuracy to the error of measurement this coincides with the optically visible extension.

The high sensitivity and accuracy in tracking exhibited by the radio telescope has made it possible for the first time to carry out observations on the occultation by the moon [7] of radio sources at a wavelength of 8 cm. Research by this method on the Crab negula has shown that in it, including the assumed location of the remains of the supernova explosion, there are no sources of low angular diameters (< 2'') of which the radio emission contribution would constitute more than 2%. The center of gravity of the nebula is displaced in a northwesterly direction relative to the center of the double star by $\Delta \alpha = -12 \pm 5''$ and $\Delta \delta = +17 \pm 5''$. This displacement is due to the presence in the northwest part of the nebula, at the location of a local formation [8] which is visible optically, of detail No. 3, the diameter of which is about 30'', and whose radio emission contribution constitutes about 8%. But the region of radio emission does not go beyond the optically visible boundaries and practically coincides with them.

By means of observations of an occultation of the radio source 3S 273 it was possible to distinguish two components in the radiation from it. It turned out that the ratio of the flux densities of these components [9] at a wavelength of 8.2 cm is 0.31.

The authors would like to express their deep gratitude to G. Ya. Gus'kov, A. D. Kuzmin, and A. E. Salomonovich for considering the article.

LITERATURE CITED

1. V. B. Shteinshleiger, G. S. Misezhnikov, and O. A. Afanas'ev, Radiotekh. i Elektron., 7:874 (1962).
2. V. B. Shteinshleiger, O. A. Afanas'ev, G. S. Misezhnikov, and Ya. I. Rozenberg, Pribory i Tekh. Éksp., 5:136 (1964).
3. B. J. Robinson, Ann. Rev. Astron. Astrophys., 2:401 (1964).
4. F. V. Bunkin and N. V. Karlov, Zh. Tekh. Fiz., 25:733 (1955).
5. L. I. Matveenko, G. S. Misezhnikov, M. M. Mukhina, and V. B. Shteinshleiger, Dokl. Akad. Nauk SSSR, 161:810 (1965).
6. V. I. Kostenko and L. I. Matveenko, Astron. Zh., 43:280 (1966).
7. L. I. Matveenko, Astron. Tsirk., No. 358, p. 5 (1966).
8. L. I. Matveenko, Astron. Tsirk., No. 343, p. 1 (1965).
9. G. B. Sholomitskii, L. I. Matveenko, and N. F. Sleptsova, Astron. Zh., 6:1135 (1965).

STATIC DESIGN OF A PARABOLIC REFLECTOR SUPPORTED AT MANY POINTS

P. D. Kalachev

Introduction

The radio telescopes with a fully steerable antenna in the form of a parabolic reflector are the basic type of radio telescope extant at the present time. The reason for this is that the steerable parabolic antenna successfully combines a number of advantageous properties, such as: ability to track the object under observation and to store the signal, capability of operation at several wavelengths simultaneously [1], relative ease and in speed in replacement of antenna feeds, and consequently easy and rapid shifting of wavelength ranges, particularly when the demountable control cabins located at the vertex of the main reflector are considered, plus some other advantages in addition.

The international conference on the design of large steerable radio antennas held in London in 1966 showed that interest in parabolic antennas has not only not slackened, but on the contrary continues to mount. At the same time, parabolic antennas require a high order of engineering art for their design, backed up by serious scientific research on dimensions of the reflector surface, on the elastic properties of the supporting structures of the reflector (load-bearing elements and ties), on shadowing of the aperture by antenna structures or spars supporting the feed system, etc. One of the most difficult problems encountered in the design of large parabolic antennas is how to achieve high mechanical rigidity in the reflector.

Deformations of the antenna reflecting surface stem from three principal causes: the effect of dead load, wind loads, and the temperature brought about by exposure to sunshine. Wind loads account for something on the order of 10% of the load caused by dead weight [2] (at the operating wind velocity $v_{op} \approx 6-12$ m/sec); in the case of highly rigid parabolic antennas, which would of course have a high relative weight. Temperature deformations can be partially or totally eliminated, as was shown in [3], and the same applies to wind loads. Deformations due to dead weight cannot be eliminated under the conditions prevailing in the earth's gravitational field. The brunt of the problem is consequently that of designing a parabolic antenna of high mechanical rigidity and thereby minimizing deformations due to the dead weight of the antenna structures.

The difficulty in achieving high reflector rigidity is that the reflector must rotate about its horizontal axis, and its elastic deformations must be as small as possible at any angle of the reflector, i.e., at any position of the reflector in space. Hence, attention must be paid to the following point. According to technological considerations, installation of the reflector panelling, check-out tests of the reflector, and any final adjustments and realignments must be carried out with the plane of the reflector aperture in the horizontal position. This means that the

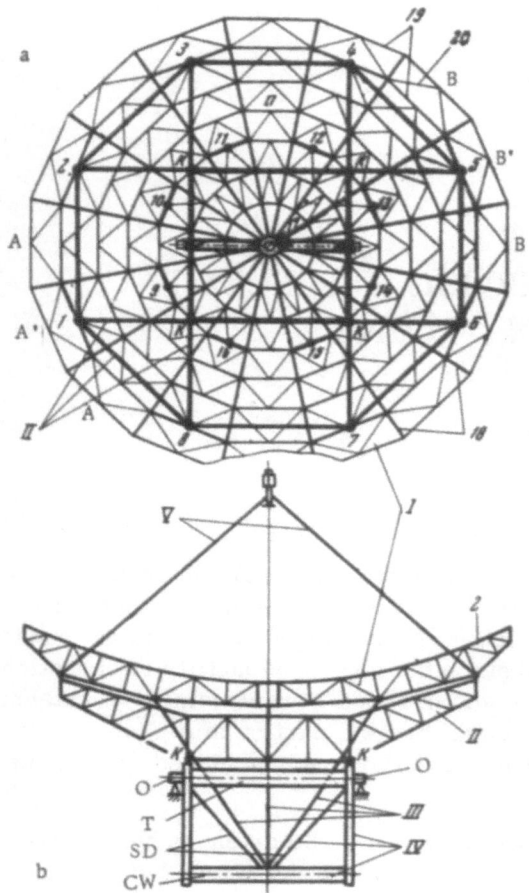

Fig. 1. Diagram showing action of forces on a parabolic reflector supported at many points. a) Plan view; b) side view along the plane of rotation of the reflector about its horizontal axis.

shape of the reflector reflecting surface will correspond most closely to the design shape when the reflector is in that position. *

It will help if we designate this particular shape of the reflector the technological shape. This reflector configuration is independent of elastic deformations due to dead weight, and for that reason can be arrived at without taking dead weight effects into account. But these deformations, compensated by adjustments in the surface, remain in, say, a latent form. If the force of gravity were eliminated with the reflector in horizontal position, the deformations would then reappear, but in the reverse direction.

The same occurs when the reflector is rotated about its horizontal axis. When the reflector aperture plane is in the vertical position, the direction of the force of gravity is parallel to the aperture plane, but in the normal direction (to the aperture plane) gravitational force is absent at that reflector position. This brings about a state at the reflector equivalent to loading by symmetric loads acting in the opposite direction. Moreover, the reflector is loaded by dead-weight forces acting parallel to the aperture plane, and the pattern of deformations becomes more complicated. The forces acting normal to the aperture plane cause symmetric deformations, while the forces acting parallel to the aperture plane are responsible for skew-symmetric deformations.

This article discusses a method for determining symmetric elastic deformations of the reflector due to dead-weight forces.

1. Force Diagram of Parabolic Reflector

Supported at Many Points

Figure 1 shows the force diagram of a parabolic reflector supported at many points, with: I) reflector; II) intermediate eight-point support structure; III) nine-bar pyramid with central bar; IV) rotation sector unit; V) eight-bar pyramid supporting the feed system; O) support trunnions; CW) counterweight beam; SD) semidiagonals; T) horizontal axis. 1, 2, . . . , 8 number the eight points of support of the reflector, formed by the eight-point intermediate support structure; 9, 10, . . . , 16, 17 number the auxiliary supports formed by the nine-bar pyramid, whose vertex rests on the counterweight beam; the letters K indicate the four nodal points of the intermediate support structure, by which this structure is joined to the rotating sectors. These

_____

*In that position, the reflector configuration will correspond to the design configuration to the level of precision of the measurements and adjustments, i.e., to within the technological errors in fabrication.

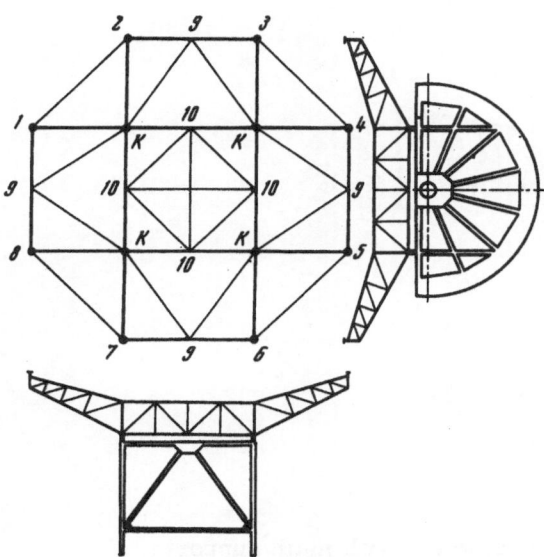

Fig. 2. Force diagram of the intermediate eight-point support structure, and the rotation sector unit.

nodal points are situated with radial symmetry about the center of the reflector. The reflector load-bearing frame is a radially symmetric space system of bars which incorporates radial members, chordal thrust members, and semi-diagonals in the lower panels, designated 18, 19, and 20, respectively.

Figure 2 shows the intermediate eight-point support structure with the rotation sectors in three projections.

Appearing in the force diagram of the intermediate support structure are: 1-4, 8-5, 2-7, 3-6, the main girders; 1-8, 4-5, 2-3, 5-7, the end girders; 1-2, 3-4, 5-6, 7-8, brace struts; K-9, semidiagonals; 10-10, tie-bars. The letters K denote the points of support of the intermediate support structure, i.e., the nodal points joining the structure to the rotation sectors.

Two rotation sectors interconnected by the counterweight beam, the horizontal axis tube, and a system of constraints joining the rim of the rotation sectors in the manner of a squirrel cage, form a rigid assembly capable of transmitting bending moments and torques applied between the support trunnions. The intermediate eight-point support structure, the nine-bar pyramid, and the rotation sectors unit combine to form the many-point support system for the parabolic reflector. In this case the support system rests on 17 points of support situated in radial symmetry about the center of the reflector.

When the reflector aperture plane is in horizontal position, the eight principal supports (1-8) formed by the intermediate support structure have the same rigidity, because of the structure's symmetry and the symmetry in loading. All the radial thrust members are situated identically with respect to the points of support, so that they all exhibit the same rigidity. Eight auxiliary points of support (9-16) formed by the side bars of the nine-bar pyramid also exhibit identical rigidity thanks to symmetry. If we now provide identical rigidity for the first and second groups of supports, as well as for support 17 formed by the central bar, then all the supports will have the same rigidity, and their displacements will not affect the shape of the reflector.

In this case we are dealing with displacements parallel to the geometrical (optical) axis of the reflector, i.e., perpendicular to the aperture plane. In addition to the vertical components, all the points of support, except the central one, will also have horizontal components of the displacement which need not be taken into account in a first approximation. This is because, for one thing, the rigidity of the reflector in the radial directions is many times greater than its rigidity in the transverse direction, i.e., along the normal to the aperture plane; for another, distortions in the reflector configuration constitute only an insignificant portion of the radial displacements, particularly in the middle portion.

2. Choice of Basic System

The design of a parabolic reflector for rigidity reduces to two principal cases: a) designing for symmetric loads, where the symmetric deformations corresponding to those loads are

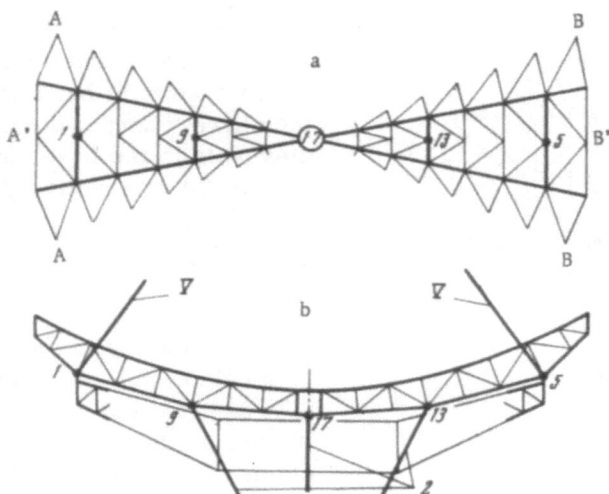

Fig. 3. Force diagram of one-fourth of the reflector with main support points and auxiliary support points singled out from remainder of reflector.

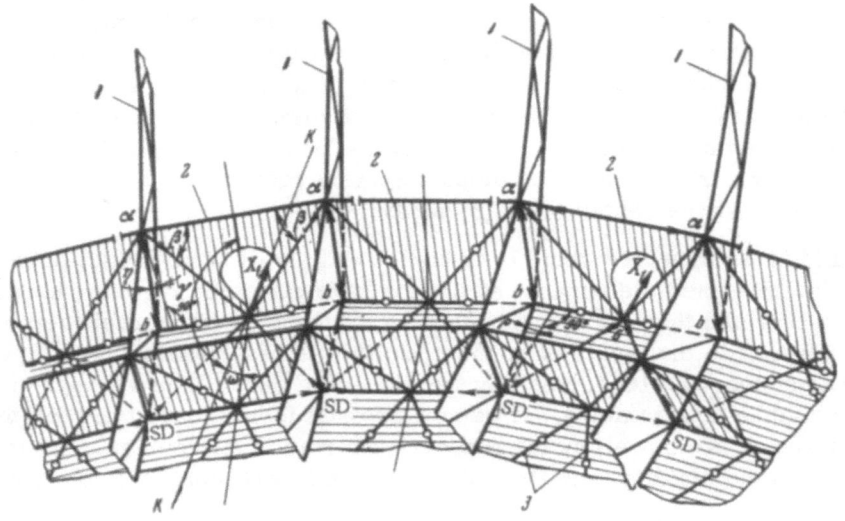

Fig. 4. Basic system of the reflector (isometric projection). 1) Radial thrust members; 2) chordal members; 3) semidiagonals in lower panels; K — K) radial direction of stress X_1.

to be determined; b) designing for skew-symmetric loads, where the skew-symmetric deformations have to be determined.

In this article, we consider the case of a symmetrically loaded reflector, corresponding to the horizontal position of the aperture plane. In that case, the symmetry of the structure and of the loading means that the stresses in all semidiagonals other than those reaching the nodal support points will be zero.

Consider, for example, nodal point n (see Fig. 1). The stresses in the semidiagonals stretching to the nodal point n must be equal in magnitude and sign, by symmetry, and accord-

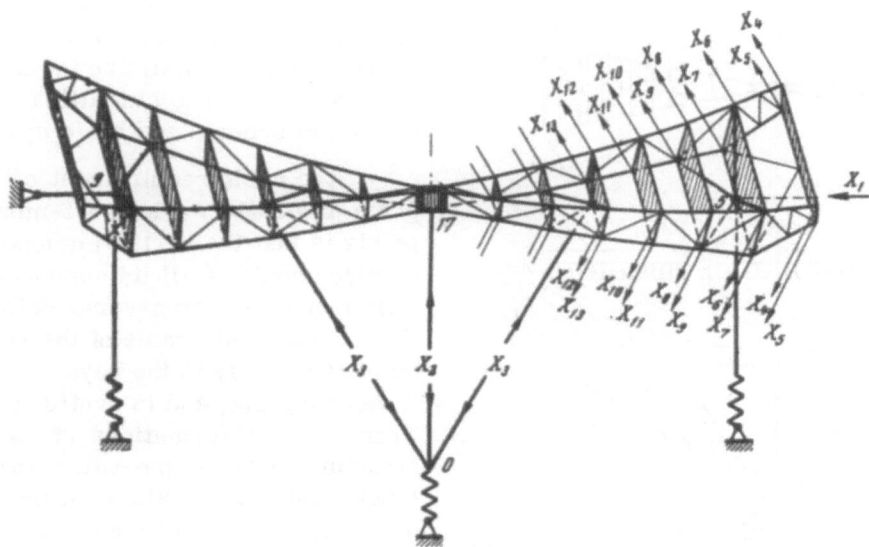

Fig. 5. Basic system for portion of reflector singled out for calculations.

ingly will yield a component normal to the chordal member. But since the chordal member is not capable of taking up stresses applied perpendicular to it, the stresses in the semidiagonals vanish. The stresses in all the respective (symmetric) members are the same. It is therefore sufficient to carry out calculations for just one fourth of the reflector, consisting of two adjacent diametrally placed thrust members mutually intersecting at the center of the reflector, and the thereto adjacent chordal members and other thrust members. This portion of the reflector, as designated in Fig. 1 by the letters A — A' — A and B — B' — B, is depicted in Fig. 3, where V are the bars (spars) of the pyramid supporting the load of the feed system; 2, the bars of the nine-bar pyramid; 1, 9, 17, 13, and 5, the supports for the reflector. The entire reflector consists of four such parts, and each of these rests on five supports. A part of the load-bearing frame of the reflector including the two main supports 5 and 6, appears in Fig. 4 in isometric projection.

As stated earlier, only those semidiagonals in the lower reflector panels which reach to the support nodal points (e.g., to points 1, 9, 13, and 5 in Fig. 3) actually do work in symmetric loading. As for the chordal (or annular) thrust members, all chordal members are in principle loaded symmetrically, but deformations, and with them stresses, in the chordal members are small in the central region of the reflector where the surface deviates only slightly from the horizontal plane, and they may be ignored in order to simplify the calculations. We can then adopt the system depicted in Fig. 5 for that portion of the reflector singled out as the main portion.

The two basic (extreme) supports imitating the intermediate support structure are elastic supports. Three auxiliary supports are formed by three bars of the nine-bar pyramid, whose vertex (O) is also fastened to the elastic support (counterweight beam).

Our redundant unknowns are: the thrust forces X_1 stemming from the effect of the intermediate support structures; forces X_2 and X_3 acting in the bars of the pyramid; forces X_4, X_5, . . . , X_{13} acting in the chordal members (hoop stresses). Here we consider the special case where the number of redundant unknowns is 13. In large parabolic reflectors, the number of chordal elements, and consequently the number of redundant unknowns, may be quite considerable. Since an increase in the number of unknowns entails a marked increase not only in the

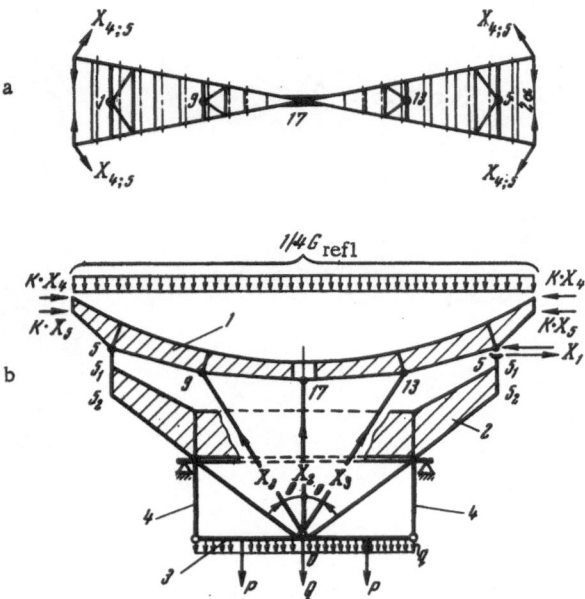

Fig. 6. Simplified basic system for the portion of the reflector singled out for calculations. a) Plan view; b) side view; K = 2 sin α. 1) Reflector; 2) intermediate support structure; 3) counterweight beam; 4) thrust members of rotation sector (spokes).

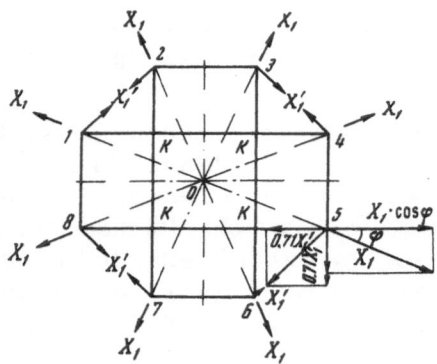

Fig. 7. Basic system for the intermediate support structure.

volume of computational labor,* but also in the number of possible errors, it becomes advantageous to minimize the number of unknowns in any valid simplification of the computational schema and basic system.

The basic requirement of the force diagram of a reflector support-mounted on many points is that the vertical components of the displacements of all its supports be the same. This means that transverse deformations of the load-bearing frame of the reflector will be possible only in the bays intervening between supports, and in cantilevered members. Transverse deformations of cantilevered structures can, in general, bring about severe longitudinal deformations of the chordal members, e.g., the members of the outer annulus, for example. But transverse deformations in the bays between supports cause only slight longitudinal deformations of the chordal (annular) members, particularly at mid-span.

Taking this circumstance into account, we can greatly simplify the basic system for the reflector. Figure 6 shows a simplified basic system for a reflector. The force coupling of that part of the reflector singled out for calculations is shown in side view, with the intermediate support structure and the counterweight beam. This system includes only five redundant unknowns: X_1, ..., X_5, where X_1 is the thrust force (effect of the intermediate structure), X_2 and X_3 are forces acting in the feed pyramid bars, X_4 and X_5 are forces acting in the chordal members of the outer annulus. In Fig. 6b, the dead weight of the one-fourth of the reflector singled out for calculations ($^1/_4 G_{refl}$) is arbitrarily shown in the form of a load distributed over the length of the projection of the diametral members. P and Q are concentrated loads on the beam due to the counterweight P and the pyramid bars Q, respectively.

The dead weight of the counterweight beam is depicted as a uniformly distributed load q. One of the features of this force diagram for the reflector and its supports is the fact that this diagram is made up of three clearly distinguishable force subunits: the load-bearing frame of the reflector with the nine-bar pyramid, the intermediate eight-point support structure, and a rotation sector with a counterweight beam. Correspondingly, the computational schema and the basic system can also be broken up into three parts: one for the portion of the reflector singled

---

*The volume of computational labor increases in proportion to roughly the square of the number of unknowns.

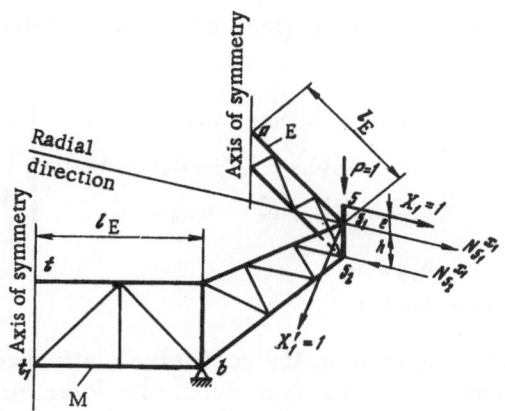

Fig. 8. Basic system for the portion of the intermediate support structure singled out for calculations. E) End semigirder; M) main semigirder.

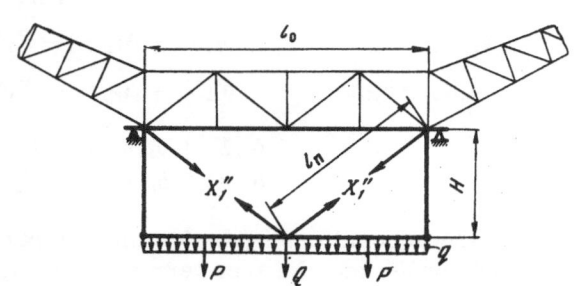

Fig. 9. Basic system for rotation sector unit joined to counterweight beam.

out in the calculations (Figs. 5 and 6), one for the intermediate support structure (Figs. 7 and 8), the third for the counterweight beam (Fig. 9).

Figure 7 depicts the basic system for the intermediate support structure (in plan view) containing one redundant unknown.

Figure 8 shows the one-fourth of the intermediate support structure (in isometric projection) corresponding to half that part of the reflector depicted in Fig. 5, i.e., to the two adjacent radial girders with the adjoining chordal members and semidiagonals.

The basic system for the unit including the rotation sectors and the counterweight beam, which also contains one redundant unknown, is depicted in Fig. 9.

The entire reflector system and its support members therefore contain a total of seven redundant unknowns, but the system divides into three independent systems, two of which contain only one unknown apiece.

3. Static Design of Reflector and Its Support

Members by the Disjointed Schema

It was noted earlier that the reflector design on the many-point suspension support (specifically, 17-point support) in question consists of three design units: the reflector with the nine-bar pyramid, the intermediate eight-point support structure, and the rotation sectors with the counterweight beam.

Static design using the disjointed schema consists in performing calculations independently on a portion of the reflector separated out in the basic system (Fig. 6), the intermediate support structure in the basic system (Figs. 7 and 8), and the unit comprising the rotation sectors joined to the counterweight beam (Fig. 9). Here the reflector system contains five redundant unknowns, while the system for the intermediate support structure and for the rotation sectors have, as stated earlier, one such unknown apiece.

Determination of the unknowns is of no particular interest to us in the case of the systems of the intermediate support structure and rotation sectors, and will not be discussed further. But the system for the reflector contains several interesting features, and their discussion will be profitable. The problem is that displacements in the basic system for the reflector depend on the elasticity of the intermediate support structure and rotation sectors.

The system of canonical equations for the basic scheme of the reflector and the nine-bar pyramid is written as follows:

$$\delta_{11}X_1 + \delta_{12}X_2 + \delta_{13}X_3 + \cdots + \delta_{15}X_5 + \Delta_{1p} = -X_1\delta'_{11} - X_2\delta'_{12} - X_3\delta'_{13} - \Delta'_{1p},$$

$$\delta_{21}X_1 + \delta_{22}X_2 + \delta_{23}X_3 + \cdots + \delta_{25}X_5 + \Delta_{2p} = -X_1\delta'_{21} - X_2(\delta'_{22} + \delta''_{22}) - X_3(\delta'_{23} + \delta''_{23}) - \Delta'_{2p} - \Delta''_{2p},$$

$$\delta_{31}X_1 + \delta_{32}X_2 + \delta_{33}X_3 + \cdots + \delta_{35}X_5 + \Delta_{3p} = -X_1\delta_{31} - X_2(\delta'_{32} + \delta''_{32}) - X_3(\delta'_{33} + \delta''_{33}) - \Delta'_{3p} - \Delta''_{3p},$$

$$\delta_{41}X_1 + \delta_{42}X_2 + \delta_{43}X_3 + \cdots + \delta_{45}X_5 + \Delta_{4p} = 0,$$

$$\delta_{51}X_1 + \delta_{52}X_2 + \delta_{53}X_3 + \cdots + \delta_{55}X_5 + \Delta_{5p} = 0. \tag{1}$$

In the general case (torsion and temperature effects neglected), the coefficients attached to the unknowns, and the free terms in the right-hand members of the equations, are determined by the familiar formulas [4]

$$\delta_{ii} = \frac{1}{E}\left(\sum \overline{X}_i^2\,\frac{l}{F}\right) + \frac{1}{E}\int \overline{M}_i^2\,\frac{dx}{I} + \frac{\mu}{G}\int \overline{Q}_i^2\,\frac{dx}{F},$$

$$\delta_{ik} = \frac{1}{E}\left(\sum \overline{X}_i\overline{X}_k\,\frac{l}{F}\right) + \frac{1}{E}\int \overline{M}_i\overline{M}_k\,\frac{dx}{I} + \frac{\mu}{G}\int \overline{Q}_i\overline{Q}_k\,\frac{dx}{F},$$

$$\Delta_{ip} = \frac{1}{E}\left(\sum \overline{X}_i N_p\,\frac{l}{F}\right) + \frac{1}{E}\int \overline{M}_i\overline{M}_p\,\frac{dx}{I} + \frac{\mu}{G}\int \overline{Q}_i\overline{Q}_p\,\frac{dx}{F},$$

where the first terms to appear are the displacements due to deformations of the bar or strut members of the system, the second and third terms are displacement due, respectively, to bending deformations and shear deformations in the truss members of the system. The right-hand members of the system of equations (1) represent the effect of the intermediate support structure and rotation sectors with counterweight beam, and are determined by the formulas

$$E\delta'_{11} = \left(\sum \overline{X}_1 N_{\Sigma x_1}\,\frac{l}{E}\right)_{\text{bar iss}} + \frac{1}{I}\int \overline{M}_1^2 dx + \frac{\mu k}{F_{\text{bar}}}\int \overline{Q}_1^2 dx,$$

where the first term is the displacement in the X_1 direction due to deformations of the bars in the intermediate support structure (bar iss), the second and third terms represent the displacement in the same direction due to bending deformations and shear deformations in the strut beam $5 - 5_2$ (see Fig. 8) forming the basic point of support for the reflector; $\overline{X}_1$ and $N_{\Sigma x_1}$ are, respectively, the stresses in the strut members of the intermediate support structure due to effects $\overline{X}_1 = 1$ in the basic system, and the total stresses in those members with the stress X'_1 in the brace strut taken into account; $\overline{M}_1$ and $\overline{Q}_1$ are, respectively, the bending moment and the transverse force in the strut beam due to the effect $X_1 = 1$; l and F are the lengths and areas of the bar (strut) transverse cross sections; I, F_{bar}, and μ are the moment of inertia, area of the transverse cross section, and coefficient of the cross section of the beam; $k = E/G$; E and G are elastic moduli of the first and the second kind.

$$E\delta'_{12} = \left(\sum \overline{X}_1 N_{\Sigma x_2}\,\frac{l}{F}\right)_{\text{bar iss}},$$

$$E\delta'_{13} = 2\delta'_{12}\cos\theta,$$

where $N_{\Sigma x_2}$ are the total stresses in strut members of the intermediate support structure due to the effect $X_2 = 1$ with the stress X'_1 in the brace strut taken into account.

Since the stresses in the truss member $5 - 5_2$ due to the effect $\overline{X}_2 = 1$ and $\overline{X}_3 = 1$ vanish, the terms containing the integrals also vanish.

$$E\Delta'_{1p} = \left(\sum \bar{X}_1 N_{\Sigma p} \frac{l}{F}\right)_{\text{bar iss}} + \int \overline{M}_1 M_{\Sigma p} \frac{dx}{I} + \mu k \int \bar{Q}_1 Q_{\Sigma p} \frac{dx}{F_{\text{bar}}}.$$

Here $N_{\Sigma p}$ are the stresses in the strut members of the intermediate support structure due to load (dead weight of system) with the stress in the brace strut taken into account; $M_{\Sigma p}$ and $Q_{\Sigma p}$ are the bending moment and the transverse force in the truss member $5 - 5_2$ with the stress X_1 in the brace strut taken into account (see Figs. 7 and 8).

$$E\delta'_{22} = \left(\sum \bar{X}_2 N_{\Sigma x_2} \frac{l}{F}\right)_{\text{bar iss}} + \left(\sum \bar{X}_2 N_{\Sigma x_2} \frac{l}{F}\right)_{\text{bar sect}} + \int \overline{M}_2 M_{\Sigma x_2} \frac{dx}{I} + \mu k \int \bar{Q}_2 Q_{\Sigma x_2} \frac{dx}{F_{\text{bar}}},$$

where the second term represents the displacement due to deformations of the strut members in the rotation sector unit due to the effect $X_2 = 1$ with the stresses in the semidiagonals X_1'' taken into account; the last two terms represent displacements due to deformations of the counterweight beam, with $M_{\Sigma x_2}$ and $Q_{\Sigma x_2}$ the bending moment and the transverse force acting in the beam due to $X_2 = 1$ with the stress in the semidiagonals X_1'' taken into account.

$$\delta'_{23} = 2\delta'_{22}\cos\theta,$$

$$E\Delta'_{2p} = \left(\sum \bar{X}_2 N_{\Sigma p} \frac{l}{F}\right)_{\text{bar iss}} + \left(\sum \bar{X}_2 N_{\Sigma p} \frac{l}{F}\right)_{\text{bar sect}} + \left(\int \overline{M}_2 M_{\Sigma p} \frac{dx}{I} + \mu k \int \bar{Q}_2 Q_{\Sigma p} \frac{dx}{F_{\text{bar}}}\right)_{\text{t.}5-5_2} +$$

$$+ \left(\int \overline{M}_2 M_{\Sigma p} \frac{dx}{I} + \mu k \int \bar{Q}_2 Q_{\Sigma p} \frac{dx}{F_{\text{bar}}}\right)_{\text{cwb}},$$

where the first two terms are the displacements due to deformations of the strut members of the intermediate support structure and rotation sector unit; the last two terms are displacements due to bending and shear deformations, respectively, of the truss members of the intermediate support structure (beam $5 - 5_2$) and rotation sector unit (counterweight beam) due to load (dead weight of the system); $N_{\Sigma p}$, $M_{\Sigma p}$, and $Q_{\Sigma p}$ are, respectively, the force factors taken with stress in the brace struts X_1' and semidiagonals X_1'' taken into account. Now,

$$\delta'_{33} = 4\delta'_{22}\cos^2\theta,$$

$$\Delta'_{3p} = 2\Delta'_{2p}\cos\theta.$$

As for δ'_{21}, δ'_{31}, δ'_{32}, these are of course equal respectively to δ'_{12}, δ'_{13}, and δ'_{23}.

The system of equations (1) can be written differently, by transferring all terms of the equations on the right to the left, and reducing similar terms:

$$(\delta_{11} + \delta'_{11})X_1 + (\delta_{12} + \delta'_{12})X_2 + (\delta_{13} + \delta'_{13})X_{13} + \cdots + \delta_{15}X_{15} + (\Delta_{1p} + \Delta'_{1p}) = 0,$$

$$(\delta_{21} + \delta'_{21})X_1 + (\delta_{22} + \delta'_{22} + \delta''_{22})X_2 + (\delta_{23} + \delta'_{23} + \delta''_{23})X_3 + \cdots \delta_{25}X_5 + (\Delta_{2p} + \Delta'_{2p}) = 0;$$

$$(\delta_{31} + \delta'_{31})X_1 + (\delta_{32} + \delta'_{32})X_2 + (\delta_{33} + \delta'_{33})X_3 + \cdots + \delta_{35}X_5 + (\Delta_{3p} + \Delta'_{3p}) = 0,$$

$$\delta_{41}X_1 + \delta_{42}X_2 + \delta_{43}X_3 + \cdots + \delta_{45}X_5 + \Delta_{4p} = 0,$$

$$\delta_{51}X_1 + \delta_{52}X_2 + \delta_{53}X_3 + \cdots + \delta_{55}X_5 + \Delta_{5p} = 0. \tag{2}$$

The notation of the system of equations (2) reflects only the (formal) fact that the sum of the displacements in directions 1, 2, . . . , 5 due to the effects of all the force factors balances out at zero, i.e., the structure is continuous. But the notation of system (2) will be needed later on, in order to demonstrate the identity of the coefficients and free terms obtained in the disjointed schema with the coefficients and free terms which we derive later on in the discussion of the static method for designing the reflector and its suspension supports as a unified system.

4. Static Design of Reflector and Its Suspension Supports as a Unified System

Here the system consists of the reflector, intermediate eight-point support structure, nine-bar pyramid, and unit combining the rotation sectors and the counterweight beam (see Figs. 6-9), and contains seven redundant unknowns: $X_1, X_2, \ldots, X_5, X_1^!$ and $X_1^{!!}$. Denoting $X_1^!$ and $X_1^{!!}$ respectively as X_6 and X_7, we now rewrite the system of canonical equations for the seven unknowns

$$\bar{\delta}_{11}X_1 + \bar{\delta}_{12}X_2 + \bar{\delta}_{13}X_3 + \bar{\delta}_{14}X_4 + \bar{\delta}_{15}X_5 + \bar{\delta}_{16}X_6 + \bar{\delta}_{17}X_7 + \bar{\Delta}_{1p} = 0,$$
$$\bar{\delta}_{21}X_1 + \bar{\delta}_{22}X_2 + \bar{\delta}_{23}X_3 + \bar{\delta}_{24}X_4 + \bar{\delta}_{25}X_5 + \bar{\delta}_{26}X_6 + \bar{\delta}_{27}X_7 + \bar{\Delta}_{2p} = 0,$$
$$\bar{\delta}_{31}X_1 + \bar{\delta}_{32}X_2 + \bar{\delta}_{33}X_3 + \bar{\delta}_{34}X_4 + \bar{\delta}_{35}X_5 + \bar{\delta}_{36}X_6 + \bar{\delta}_{37}X_7 + \bar{\Delta}_{3p} = 0,$$
$$\bar{\delta}_{41}X_1 + \bar{\delta}_{42}X_2 + \bar{\delta}_{43}X_3 + \bar{\delta}_{44}X_4 + \bar{\delta}_{45}X_5 + \bar{\delta}_{46}X_6 + \bar{\delta}_{47}X_7 + \bar{\Delta}_{4p} = 0, \qquad (3)$$
$$\bar{\delta}_{51}X_1 + \bar{\delta}_{52}X_2 + \bar{\delta}_{53}X_3 + \bar{\delta}_{54}X_4 + \bar{\delta}_{55}X_5 + \bar{\delta}_{56}X_6 + \bar{\delta}_{57}X_7 + \bar{\Delta}_{5p} = 0,$$
$$\bar{\delta}_{61}X_1 + \bar{\delta}_{62}X_2 + \bar{\delta}_{63}X_3 + \bar{\delta}_{64}X_4 + \bar{\delta}_{65}X_5 + \bar{\delta}_{66}X_6 + \bar{\delta}_{67}X_7 + \bar{\Delta}_{6p} = 0,$$
$$\bar{\delta}_{71}X_1 + \bar{\delta}_{72}X_2 + \bar{\delta}_{73}X_3 + \bar{\delta}_{74}X_4 + \bar{\delta}_{75}X_5 + \bar{\delta}_{76}X_6 + \bar{\delta}_{77}X_7 + \bar{\Delta}_{7p} = 0.$$

Here the coefficients attached to the unknowns, and the free terms, are generally different from the coefficients and free terms of system (1), and are therefore written with bars above.

The coefficients and the free terms in system (3) can be expressed in terms of the coefficients and free terms in system (1) in the following manner:

$$\bar{\delta}_{11} = 2\left(\sum \bar{X}_1^2 \frac{l}{EF}\right)_{r.g} + \left(\sum \bar{X}_1^2 \frac{l}{EF}\right)_{bar\,iss} + \int \bar{M}_1^2 \frac{dx}{EI} + \mu \int \bar{Q}^2 \frac{dx}{GF_{bar}}, \qquad (4)$$

where the first term represents the displacement due to deformation of the struts in two adjoining radial girders in the reflector (see Figs. 5 and 6), as well as the bars of the nine-bar pyramid; the second term represents the displacement due to deformations of the bars (struts) in the intermediate support structure, and the last two terms account for displacement due to deformation of truss members in the intermediate support structure. But

$$2\left(\sum \bar{X}_1^2 \frac{l}{EF}\right)_{r.g} = \delta_{11}, \qquad (5)$$

$$\left(\sum \bar{X}_1^2 \frac{l}{EF}\right)_{bar\,iss} + \int \bar{M}_1^2 \frac{dx}{EI} + \mu \int \bar{Q}_1^2 \frac{dx}{EF_{bar}} = \delta_{11}' + \frac{\bar{\delta}_{16}^2}{\bar{\delta}_{66}}. \qquad (6)$$

Equation (6) can be proved as follows. We represent the displacement of point 5 on the intermediate support structure (see Fig. 8) in the X_1 direction as

$$\delta_{11}' = \left(\sum \bar{X}_1^2 \frac{l}{EF}\right)_{bar\,iss} + \int \bar{M}_1^2 \frac{dx}{EI} + \mu \int \bar{Q}_1^2 \frac{dx}{GF_{bar}} - \left[\left(\sum \bar{X}_1 \bar{X}_1' \frac{l}{EF}\right)_{bar\,iss}\right] X_1'. \qquad (7)$$

Here the first three terms are the displacement of point 5, while ignoring the effect of the brace strut in the intermediate support structure; they are followed by a binomial term (in square brackets) multiplied by X_1' (the stress in the brace strut) giving the displacement of point 5 due to the effect of X_1'.

The stress in the brace strut is in turn determined from the basic system for the intermediate support structure (Figs. 7 and 8) by using the Maxwell — Mohr formula [4]

$$X_1' = \frac{-\Delta_{1p}^{r.q}}{\delta_{11}^{r.q}} = \frac{\left(\sum \overline{X}_1 \overline{X}_1' \frac{l}{EF}\right)_{\text{bar iss}}}{\left[\sum (\overline{X}_1')^2 \frac{l}{EF}\right]_{\text{bar iss}}}. \tag{8}$$

But

$$\left.\begin{aligned}
\left(\sum \overline{X}_1 \overline{X}_1' \frac{l}{EF}\right)_{\text{bar iss}} = \left(\sum \overline{X}_1 \overline{X}_6 \frac{l}{EF}\right)_{\text{bar iss}} = \overline{\delta}_{16} \\[2mm]
\left[\sum (\overline{X}_1')^2 \frac{l}{EF}\right]_{\text{bar iss}} = \overline{\delta}_{66}.
\end{aligned}\right\} \tag{9}$$

and

Accordingly, $\overline{X}_1' = \overline{\delta}_{16}/\overline{\delta}_{66}$.

Substitution of the value of X_1' from Eq. (9) into Eq. (7), with Eq. (8) taken into account, yields

$$\delta_{11}' = \left(\sum \overline{X}_1^2 \frac{l}{EF}\right)_{\text{bar iss}} + \int \overline{M}_1^2 \frac{dx}{EI} + \mu \int \overline{Q}_1^2 \frac{dx}{GF_{\text{bar}}} - \frac{\overline{\delta}_{16}^2}{\overline{\delta}_{66}}$$

or

$$\left(\sum \overline{X}_1^2 \frac{l}{EF}\right)_{\text{bar iss}} + \int \overline{M}_1^2 \frac{dx}{EI} + \mu \int \overline{Q}_1^2 \frac{dx}{GF_{\text{bar}}} = \delta_{11}' + \frac{\overline{\delta}_{16}^2}{\overline{\delta}_{66}}.$$

From Eqs. (4)-(6) we get

$$\overline{\delta}_{11} = \delta_{11} + \delta_{11} + \frac{\overline{\delta}_{16}^2}{\overline{\delta}_{66}}. \tag{10}$$

Similarly,

$$\overline{\delta}_{12} = 2\left(\sum X_1 X_2 \frac{l}{EF}\right)_{\text{r.g}} + \left(\sum \overline{X}_1 \overline{X}_2 \frac{l}{EF}\right)_{\text{bar iss}}, \tag{11}$$

where

$$2\left(\sum \overline{X}_1 \overline{X}_2 \frac{l}{EF}\right)_{\text{r.g}} = \delta_{12} \tag{12}$$

and

$$\delta_{12}' = \left(\sum \overline{X}_1 \overline{X}_2 \frac{l}{EF}\right)_{\text{bar iss}} - \left(\sum \overline{X}_1 \overline{X}_1' \frac{l}{EF}\right)_{\text{bar iss}} X_1', \tag{13}$$

where

$$X_1' = \frac{\left(\sum \overline{X}_2 \overline{X}_1' \frac{l}{EF}\right)_{\text{bar iss}}}{\left[\sum (\overline{X}_1')^2 \frac{l}{EF}\right]_{\text{bar iss}}} = \frac{\overline{\delta}_{76}}{\overline{\delta}_{66}}, \tag{14}$$

$$\left(\sum \overline{X}_1 \overline{X}_1' \frac{l}{EF}\right)_{\text{bar iss}} = \overline{\delta}_{16}. \tag{15}$$

Substitution of the corresponding values from Eqs. (14) and (15) into Eq. (13) yields

$$\delta_{12}' = \left(\sum \overline{X}_1 \overline{X}_2 \frac{l}{EF}\right)_{\text{bar iss}} - \overline{\delta}_{16} \frac{\overline{\delta}_{26}}{\overline{\delta}_{66}}. \tag{16}$$

And hence

$$\left(\sum \overline{X}_1 \overline{X}_2 \frac{l}{EF}\right)_{\text{bar iss}} = \delta_{12}' - \frac{\overline{\delta}_{16} \overline{\delta}_{26}}{\overline{\delta}_{66}}. \tag{17}$$

Substitution of Eqs. (12) and (17) into Eq. (11) yields

$$\overline{\delta}_{12} = \delta_{12} + \delta_{12}' + \frac{\overline{\delta}_{16} \overline{\delta}_{26}}{\overline{\delta}_{66}}. \tag{18}$$

And, similarly,

$$\overline{\delta}_{13} = \delta_{13} + \delta_{13}' + \frac{\overline{\delta}_{16} \overline{\delta}_{36}}{\overline{\delta}_{66}}, \tag{19}$$

$$\overline{\delta}_{14} = \delta_{14}, \quad \overline{\delta}_{15} = \delta_{15}. \tag{20}$$

The displacement due to the effect $X_7 = 1$, i.e., $X_7' = 1$, in the X_1 direction will be zero,

$$\delta_{17} = 0,$$

since the stresses in the radial truss members and in the intermediate support structure due to the effect $X_7 = 1$ will be zero, and the stresses in the members of the rotation sector unit due to the effect $X_1 = 1$ will be zero.

Similarly,

$$\overline{\Delta}_{1p} = \Delta_{1p} + \Delta_{1p}' + \frac{\overline{\delta}_{16} \overline{\Delta}_{1p}}{\overline{\delta}_{66}},$$
$$\overline{\delta}_{21} = \overline{\delta}_{12} = \delta_{12} + \delta_{12}' + \frac{\overline{\delta}_{16} \overline{\delta}_{26}}{\overline{\delta}_{66}}, \tag{21}$$

$$\delta_{22} = 2\left(\sum \overline{X}_2^2 \frac{l}{EF}\right)_{\text{r.g}} + \left(\sum \overline{X}_2^2 \frac{l}{EF}\right)_{\text{bar iss}} + \left(\int \overline{M}_2^2 \frac{dx}{EI} + \mu \int \overline{Q}_2^2 \frac{dx}{GF_{\text{bar}}}\right)_{\text{cwb}}, \tag{22}$$

where

$$2\left(\sum \overline{X}_2^2 \frac{l}{EF}\right)_{\text{r.g}} = \delta_{22}, \tag{23}$$

$$\left(\sum \overline{X}_2^2 \frac{l}{EF}\right)_{\text{bar iss}} = \delta_{22}' + \frac{\overline{\delta}_{26}^2}{\overline{\delta}_{66}}, \tag{24}$$

$$\left(\int \overline{M}_2^2 \frac{dx}{EI} + \mu \int \overline{Q}_2^2 \frac{dx}{GF_{\text{bar}}}\right)_{\text{cwb}} = \delta_{22}'' + \frac{\overline{\delta}_{27}^2}{\overline{\delta}_{77}}. \tag{25}$$

Equations (24) and (25), representing displacements due to deformations of the intermediate support structure and counterweight beam of the rotation sectors, can be expressed by the following relationship:

$$\delta'_{22} = \left(\sum \overline{X}_2^2 \frac{l}{EF}\right)_{\text{bar iss}} - \left(\sum \overline{X}_2 \overline{X}'_1 \frac{l}{EF}\right)_{\text{bar iss}} X'_1,$$

where

$$X'_1 = \frac{\left(\sum \overline{X}_2 \overline{X}'_1 \frac{l}{EF}\right)_{\text{bar iss}}}{\left[\sum (\overline{X}'_2)^2 \frac{l}{EF}\right]_{\text{bar iss}}} = \frac{\left(\sum \overline{X}_2 \overline{X}_6 \frac{l}{EF}\right)_{\text{bar iss}}}{\left(\sum \overline{X}_6^2 \frac{l}{EF}\right)_{\text{bar iss}}} = \frac{\overline{\delta}_{26}}{\overline{\delta}_{66}},$$

$$\delta'_{22} = \left(\sum \overline{X}_2^2 \frac{l}{EF}\right)_{\text{bar iss}} - \frac{\overline{\delta}_{26}^2}{\overline{\delta}_{66}};$$

and hence

$$\left(\sum \overline{X}_2^2 \frac{l}{EF}\right)_{\text{bar iss}} = \delta'_{22} + \frac{\overline{\delta}_{26}^2}{\overline{\delta}_{66}}.$$

Similarly, for Eq. (25),

$$\delta''_{22} = \left(\int \frac{\overline{M}_2^2}{EI} dx + \mu \int \frac{\overline{Q}_2^2}{GF} dx\right)_{\text{cwb}} - \left(\int \frac{\overline{M}_2 \overline{M}'_1}{EI} dx + \mu \int \frac{\overline{Q}_2 \overline{Q}''_1}{GE} dx\right)_{\text{cwb}} X''_1, \tag{26}$$

where

$$X'' = \frac{\left(\int \frac{\overline{M}_2 \overline{M}''_1}{EI} dx + \mu \int \frac{\overline{Q}_2 \overline{Q}''_1}{GF} dx\right)_{\text{cwb}}}{\left(\int (\overline{M}''_1)^2 \frac{dx}{EI} + \mu \int \frac{\overline{Q}_2 \overline{Q}''_1}{GF} dx\right)_{\text{cwb}}} = \frac{\overline{\delta}_{27}}{\overline{\delta}_{77}}. \tag{27}$$

From Eqs. (26) and (27), we have

$$\delta''_{22} = \left(\int \frac{\overline{M}_2^2}{EI} dx + \mu \int \frac{\overline{Q}_2^2}{GF} dx\right)_{\text{cwb}} - \frac{\overline{\delta}_{27}^2}{\overline{\delta}_{77}};$$

and hence

$$\left(\int \frac{\overline{M}_2^2}{EI} dx + \mu \int \frac{\overline{Q}_2^2}{GF} dx\right)_{\text{cwb}} = \delta''_{22} + \frac{\overline{\delta}_{27}^2}{\overline{\delta}_{77}}.$$

Substitution of the corresponding values from Eqs. (23)–(25) into Eq. (22) yields

$$\overline{\delta}_{22} = \delta_{22} + \delta'_{22} + \frac{\overline{\delta}_{26}^2}{\overline{\delta}_{66}} + \delta''_{22} + \frac{\overline{\delta}_{27}^2}{\overline{\delta}_{77}}. \tag{28}$$

Again, similarly,

$$\overline{\delta}_{23} = \delta_{23} + \delta''_{23} + \frac{\overline{\delta}_{26} \overline{\delta}_{36}}{\overline{\delta}_{66}} + \delta'_{23} + \frac{\overline{\delta}_{27} \overline{\delta}_{37}}{\overline{\delta}_{77}}, \tag{29}$$

$$\overline{\Delta}_{2p} = \Delta_{2p} + \Delta''_{2p} + \frac{\overline{\delta}_{26}}{\overline{\delta}_{66}} \overline{\Delta}_{6p} + \Delta'_{2p} + \frac{\overline{\delta}_{27}}{\overline{\delta}_{77}} \overline{\Delta}_{7p}, \tag{30}$$

$$\left. \begin{aligned} &\overline{\delta}_{31} = \overline{\delta}_{13}, \quad \overline{\varrho}_{32} = \overline{\delta}_{23}, \\ &\overline{\delta}_{33} = \delta_{33} + \delta'_{33} + \frac{\overline{\delta}_{36}^2}{\overline{\delta}_{66}} + \frac{\overline{\delta}_{37}^2}{\overline{\delta}_{77}} + \delta''_{33}, \end{aligned} \right\} \tag{31}$$

$$\overline{\Delta}_{3p} = \Delta_{3p} + \Delta'_{5p} + \frac{\overline{\delta}_{36}}{\overline{\delta}_{66}} \overline{\Delta}_{6p} + \Delta''_{3p} + \frac{\overline{\delta}_{37}}{\overline{\delta}_{77}} \Delta_{7p}. \tag{32}$$

Note here that

$$\overline{\delta}_{46} = \overline{\delta}_{47} = \overline{\delta}_{56} = \overline{\delta}_{57} = 0,$$

since the stresses in the members of the intermediate support structure and in the counter-weight beam due to the effects $X_4 = 1$ and $X_5 = 1$ are zero, and the stresses in the radial truss members due to the effects $\overline{X}_6 = 1$ and $\overline{X}_7 = 1$ are also zero.

Hence, the system of equations (3) becomes

$$\begin{aligned}
&\overline{\delta}_{11}X_1 + \overline{\delta}_{12}X_2 + \overline{\delta}_{13}X_3 + \overline{\delta}_{14}X_4 + \overline{\delta}_{15}X_5 + \overline{\delta}_{16}X_6 + 0 + \overline{\Delta}_{1p} = 0, \\
&\overline{\delta}_{21}X_1 + \overline{\delta}_{22}X_2 + \overline{\delta}_{23}X_3 + \overline{\delta}_{24}X_4 + \overline{\delta}_{25}X_5 + \overline{\delta}_{26}X_6 + \overline{\delta}_{27}X_7 + \overline{\Delta}_{2p} = 0, \\
&\overline{\delta}_{31}X_1 + \overline{\delta}_{32}X_2 + \overline{\delta}_{33}X_3 + \overline{\delta}_{34}X_4 + \overline{\delta}_{35}X_5 + \overline{\delta}_{36}X_6 + \overline{\delta}_{37}X_7 + \overline{\Delta}_{3p} = 0, \\
&\overline{\delta}_{41}X_1 + \overline{\delta}_{42}X_2 + \overline{\delta}_{43}X_3 + \overline{\delta}_{44}X_4 + \overline{\delta}_{45}X_5 + 0 + 0 + \overline{\Delta}_{4p} = 0, \\
&\overline{\delta}_{51}X_1 + \overline{\delta}_{52}X_2 + \overline{\delta}_{53}X_3 + \overline{\delta}_{54}X_4 + \overline{\delta}_{55}X_5 + 0 + 0 + \overline{\Delta}_{5p} = 0; \\
&\overline{\delta}_{61}X_1 + \overline{\delta}_{62}X_2 + \overline{\delta}_{63}X_3 + 0 + 0 + \overline{\delta}_{66}X_6 + 0 + \overline{\Delta}_{6p} = 0, \\
&0 + \overline{\delta}_{72}X_2 + \overline{\delta}_{73}X_3 + 0 + 0 + 0 + \overline{\delta}_{77}X_7 + \overline{\Delta}_{7p} = 0.
\end{aligned} \tag{33}$$

It is clear from Eq. (33) that the resulting system of seven equations in seven unknowns can be converted into a system of five equations in five unknowns.

Utilizing system (33), we now express X_6 and X_7 in terms of X_1, X_2, and X_3:

$$X_6 = \frac{\overline{\delta}_{61}X_1 + \overline{\delta}_{62}X_2 + \overline{\delta}_{63}X_3 + \overline{\Delta}_{6p}}{\overline{\delta}_{66}},$$

$$X_7 = \frac{\overline{\delta}_{72}X_2 + \overline{\delta}_{73}X_3 + \overline{\Delta}_{7p}}{\overline{\delta}_{77}}.$$

Substituting the values of X_6 and X_7 in the first three equations of system (33), and reducing similar terms, we end up with a system of five equations in five unknowns:

$$\begin{aligned}
&\overline{\overline{\delta}}_{11}X_1 + \overline{\overline{\delta}}_{12}X_2 + \overline{\overline{\delta}}_{13}X_3 + \overline{\overline{\delta}}_{14}X_4 + \overline{\overline{\delta}}_{15}X_5 + \overline{\overline{\Delta}}_{1p} = 0, \\
&\overline{\overline{\delta}}_{21}X_1 + \overline{\overline{\delta}}_{22}X_2 + \overline{\overline{\delta}}_{23}X_3 + \overline{\overline{\delta}}_{24}X_4 + \overline{\overline{\delta}}_{25}X_5 + \overline{\overline{\Delta}}_{2p} = 0, \\
&\overline{\overline{\delta}}_{31}X_1 + \overline{\overline{\delta}}_{32}X_2 + \overline{\overline{\delta}}_{33}X_3 + \overline{\overline{\delta}}_{34}X_4 + \overline{\overline{\delta}}_{35}X_5 + \overline{\overline{\Delta}}_{3p} = 0, \\
&\overline{\overline{\delta}}_{41}X_1 + \overline{\overline{\delta}}_{42}X_2 + \overline{\overline{\delta}}_{43}X_3 + \overline{\overline{\delta}}_{44}X_4 + \overline{\overline{\delta}}_{45}X_5 + \overline{\overline{\Delta}}_{4p} = 0, \\
&\overline{\overline{\delta}}_{51}X_1 + \overline{\overline{\delta}}_{52}X_2 + \overline{\overline{\delta}}_{53}X_3 + \overline{\overline{\delta}}_{54}X_4 + \overline{\overline{\delta}}_{55}X_5 + \overline{\overline{\Delta}}_{5p} = 0,
\end{aligned} \tag{34}$$

where

$$\begin{aligned}
&\overline{\overline{\delta}}_{11} = \overline{\delta}_{11} - \frac{\overline{\delta}_{16}^2}{\overline{\delta}_{66}}, \quad \overline{\overline{\delta}}_{12} = \overline{\delta}_{12} - \frac{\overline{\delta}_{16}\overline{\delta}_{26}}{\overline{\delta}_{66}}, \\
&\overline{\overline{\delta}}_{13} = \overline{\delta}_{13} - \frac{\overline{\delta}_{16}\overline{\delta}_{36}}{\overline{\delta}_{66}}, \quad \overline{\overline{\Delta}}_{1p} = \overline{\Delta}_{1p} - \frac{\overline{\delta}_{16}}{\overline{\delta}_{66}} \overline{\Delta}_{6p}, \\
&\overline{\overline{\delta}}_{22} = \overline{\delta}_{22} - \frac{\overline{\delta}_{26}^2}{\overline{\delta}_{66}} - \frac{\overline{\delta}_{27}^2}{\overline{\delta}_{77}}, \\
&\overline{\overline{\delta}}_{23} = \overline{\delta}_{23} - \frac{\overline{\delta}_{26}\overline{\delta}_{36}}{\overline{\delta}_{66}} - \frac{\overline{\delta}_{27}\overline{\delta}_{37}}{\overline{\delta}_{77}},
\end{aligned} \tag{35}$$

$$\bar{\bar{\Delta}}_{2p} = \bar{\Delta}_{2p} - \frac{\bar{\delta}_{26}}{\bar{\delta}_{66}} \bar{\Delta}_{6p} - \frac{\bar{\delta}_{27}}{\bar{\delta}_{77}} \bar{\Delta}_{7p},$$ (35)

$$\bar{\bar{\delta}}_{33} = \bar{\delta}_{33} - \frac{\bar{\delta}_{36}^2}{\bar{\delta}_{66}} - \frac{\bar{\delta}_{37}^2}{\bar{\delta}_{77}},$$

$$\bar{\bar{\Delta}}_{3p} = \bar{\Delta}_{3p} - \frac{\bar{\delta}_{36}}{\bar{\delta}_{66}} \bar{\Delta}_{6p} - \frac{\bar{\delta}_{37}}{\bar{\delta}_{77}} \bar{\Delta}_{7p}.$$

Substituting the values of $\bar{\delta}_{11}, \bar{\delta}_{12}, \ldots, \bar{\delta}_{ik}, \bar{\Delta}_{ik}$ from Eqs. (10), (18), (19), (21), (28)–(32), into Eqs. (35), we get

$$\bar{\bar{\delta}}_{11} = \delta_{11} + \delta_{11}' + \frac{\bar{\delta}_{16}^2}{\bar{\delta}_{66}} - \frac{\bar{\delta}_{16}^2}{\bar{\delta}_{66}} = \delta_{11} + \delta_{11}',$$

$$\bar{\bar{\delta}}_{12} = \delta_{12} + \delta_{12}' + \frac{\bar{\delta}_{16}\bar{\delta}_{26}}{\bar{\delta}_{66}} - \frac{\bar{\delta}_{16}\bar{\delta}_{26}}{\bar{\delta}_{66}} = \delta_{12} + \delta_{12}',$$

$$\bar{\bar{\delta}}_{13} = \delta_{13} + \delta_{13}' + \frac{\bar{\delta}_{16}\bar{\delta}_{36}}{\bar{\delta}_{66}} - \frac{\bar{\delta}_{16}\bar{\delta}_{36}}{\bar{\delta}_{66}} = \delta_{13} + \delta_{13}',$$

$$\bar{\bar{\Delta}}_{1p} = \Delta_{1p} \Delta_{1p}' + \frac{\bar{\delta}_{16}\bar{\Delta}_{6p}}{\bar{\delta}_{66}} - \frac{\bar{\delta}_{16}\bar{\Delta}_{6p}}{\bar{\delta}_{66}} = \Delta_{1p} + \Delta_{1p}',$$

$$\bar{\bar{\delta}}_{22} = \delta_{22} + \delta_{22}' + \frac{\bar{\delta}_{26}^2}{\bar{\delta}_{66}} + \delta_{22}'' + \frac{\bar{\delta}_{27}^2}{\bar{\delta}_{77}} - \frac{\bar{\delta}_{26}^2}{\bar{\delta}_{66}} - \frac{\bar{\delta}_{27}^2}{\bar{\delta}_{77}} = \delta_{22} + \delta_{22}' + \delta_{22}'',$$ (36)

$$\bar{\bar{\delta}}_{23} = \delta_{23} + \delta_{23}' + \frac{\bar{\delta}_{26}\bar{\delta}_{36}}{\bar{\delta}_{66}} + \delta_{23}'' + \frac{\bar{\delta}_{27}\bar{\delta}_{37}}{\bar{\delta}_{77}} - \frac{\bar{\delta}_{26}\bar{\delta}_{36}}{\bar{\delta}_{66}} - \frac{\bar{\delta}_{27}\bar{\delta}_{37}}{\bar{\delta}_{77}} = \delta_{23} + \delta_{23}' + \delta_{23}'',$$

$$\bar{\bar{\delta}}_{33} = \delta_{33} + \delta_{33}' + \frac{\bar{\delta}_{36}^2}{\bar{\delta}_{66}} + \frac{\bar{\delta}_{37}^2}{\bar{\delta}_{77}} + \delta_{33}'' - \frac{\bar{\delta}_{36}^2}{\bar{\delta}_{66}} - \frac{\bar{\delta}_{37}^2}{\bar{\delta}_{77}} = \delta_{33} + \delta_{33}' + \delta_{33}'',$$

$$\bar{\bar{\Delta}}_{2p} = \Delta_{2p} + \Delta_{2p}' + \frac{\bar{\delta}_{26}}{\bar{\delta}_{66}} \bar{\Delta}_{6p} + \Delta_{2p}'' + \frac{\bar{\delta}_{27}}{\bar{\delta}_{77}} \bar{\Delta}_{7p} - \frac{\bar{\delta}_{26}\bar{\Delta}_{6p}}{\bar{\delta}_{66}} - \frac{\bar{\delta}_{27}}{\bar{\delta}_{77}} \bar{\Delta}_{7p} = \Delta_{2p} + \Delta_{2p}' + \Delta_{2p}'',$$

$$\bar{\bar{\Delta}}_{3p} = \Delta_{3p} + \Delta_{3p}' + \frac{\bar{\delta}_{36}}{\bar{\delta}_{66}} \bar{\Delta}_{6p} + \Delta_{3p}'' + \frac{\bar{\delta}_{37}}{\bar{\delta}_{77}} \bar{\Delta}_{7p} - \frac{\bar{\delta}_{36}}{\bar{\delta}_{66}} \bar{\Delta}_{6p} - \frac{\bar{\delta}_{37}}{\bar{\delta}_{77}} \bar{\Delta}_{7p} = \Delta_{3p} + \Delta_{3p}' + \Delta_{3p}''.$$

The resulting coefficients and free terms coincide with the respective coefficients and free terms of the system of equations (1).

This means that our system: reflector, intermediate support structure, and rotation sector unit, containing seven redundant unknowns, can be solved by using the system of equations in five unknowns and using the two equations each containing one unknown.

After determining the redundant unknowns, we can do the remaining static design calculations in the usual manner.

1. We find the computed (total) stresses in all the members of the system, using the formulas

$$N_{\text{comp}} = N_\Sigma = \sum_{i=1}^{i=5} \bar{X}_i X_i + N_p,$$

$$M_{\text{comp}} = M_\Sigma = \Sigma \bar{M}_i X_i + M_p,$$ (37)

$$Q_{\text{comp}} = Q_\Sigma = \Sigma \bar{Q}_i X_i + Q_p,$$

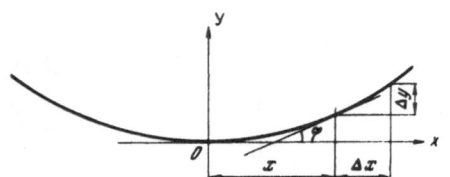

Fig. 10. Diagram of profile of reflector reflecting surface.

where N_Σ, M_Σ, Q_Σ are the total stresses applied to the system's truss members by stress in its strut members, by bending moments, and by transverse forces, due to the effects X_1, X_2, ... , X_5; N_p, M_p, Q_p are the corresponding stresses due to load (dead weight of the system).

2. A static and a kinematic check is done on the computed stresses, with the criterion for correct solution of the problem being that all internal and external forces acting on any unit of the system sum to zero, and displacements in any cross section of a thrust member of the structure balance out to zero (continuity condition of the structure).

3. The vertical components of the displacements of the eight main supports and nine auxiliary supports of the reflector are determined. If the vertical components of the main supports and auxiliary supports are identical, then the displacements in the midspan and cantilevered portions of the reflector are determined. If the vertical components of the displacements of supports are not identical, we achieve identical displacements by strengthening or weakening cross sections of the counterweight beam semidiagonals and of the bars in the nine-bar pyramid. As calculations using the schema discussed show, it is sufficient to carry out six to eight recalculations with appropriately modified cross sections in order to arrive at a satisfactory result.

The displacements are determined in the usual manner:

$$\delta_i = \frac{1}{E} \Sigma \overline{N}_1 N_\Sigma \frac{l}{F} + \frac{1}{E} \int \overline{M}_1 M_\Sigma \frac{dx}{I} + \frac{\mu}{G} \int \overline{Q}_1 Q_\Sigma \frac{dx}{F_{bar}} \quad , \tag{38}$$

where the first term is the displacement due to deformations of the strut members of the system, and the last two terms are displacements due to bending and shear deformations of the truss members of the system; $\overline{N}_1$, $\overline{M}_1$, and $\overline{Q}_1$ are stresses in the strut members and, respectively, bending moments and transverse forces in the truss members of the system due to unit load applied at the point and in the direction of the unknown displacement; N_Σ, M_Σ, and Q_Σ are the computed stresses, bending moments, and transverse forces in the respective members.

There are also horizontal components of the displacements in the main and auxiliary supports of the reflector, in addition to the central support. They correspond to some additional vertical displacements (or displacements normal to the reflector profile) which distort the reflecting surface of the reflector.

Consider the possible magnitudes of these additional displacements. It is clear in Fig. 10 that corresponding to the additional horizontal displacement Δx we have a transverse displacement, i.e., a distortion of the reflector profile Δy equal to

$$\Delta y = \Delta x \tan \varphi,$$

where $\tan \varphi = y'$. For the reflector parabola, we have

$$y = \frac{x^2}{4F}, \quad y' = \frac{x}{2F},$$

where F is the focal distance. Usually, $F = (0.33-0.53)D_{refl}$, where D_{refl} is the reflector diameter.

Assuming $F = 0.43D_{refl}$, and determining the displacements at the points of auxiliary and main supports, we obtain:

a) for the radius at which the auxiliary supports are placed, roughly at $0.2D_{refl}$

$$y'_{aux\ sup} = \frac{0.2D_{refl}}{4 \cdot 0.4D_{refl}} = 0.12,$$

$$\Delta y_{aux\ sup} = 0.12\Delta x,$$

i.e., the distortion of the reflector profile due to horizontal displacements of the auxiliary supports is only about 12% of the horizontal displacement;

b) at the radius at which the main supports are placed, roughly at $0.4D_{refl}$:

$$y'_{main\ sup} \approx 0.24,$$

$$\Delta y_{main\ sup} = 0.24\Delta x,$$

i.e., distortion of the reflector profile due to horizontal displacements of the main supports amounts to about 24% of the horizontal displacement.

The magnitudes of the horizontal displacements of the supports themselves prove to be insignificant, since the rigidity of the reflector in the vertical directions, i.e., in the planes parallel to the reflector aperture plane, is several times greater than the rigidity in the transverse direction.

As calculations have shown [5], the horizontal displacements of the auxiliary points of support amount to about 10% of the vertical displacements, while the horizontal displacements of the main points of support amount to about 25% of the corresponding vertical displacements.

Hence, the horizontal displacements of the supports do not seriously affect distortion of the reflector profile, and can be disregarded in a first approximation.

LITERATURE CITED

1. I. V. Vavilova, G. N. Galimov, P. D. Kalachev, A. M. Karachun, A. D. Kuz'min, B. N. Losovskii, and A. E. Salomonovich, Vopr. Radioelektron., Seriya Obshchetekhn. [Advances in Radio Electronics, General Engineering Series], No. 1, p. 14 (1964).
2. P. D. Kalachev, in: Radio Telescopes (D. V. Skobel'tsyn, ed.), Consultants Bureau, New York (1966), p. 35.
3. Yu. L. Shakhbazyan, Izv. GAO (State Astron. Obs. Bulletin), Vol. 23, Issue 3, No. 172, p. 180 (1964).
4. I. M. Rabinovich, A Course on the Structural Mechanics of Strut Systems, Part 2 [in Russian], Gosstroiizdat (1940).
5. P. D. Kalachev, V. P. Nazarov, V. Ya. Chashnikov, and A. A. Parshchikov, this volume, p. 53.

RIGIDITY OF CANTILEVER MOUNTING
OF PARABOLIC MIRRORS

P. D. Kalachev

The method of mounting a fully rotational parabolic mirror (parabolic antenna) on the rotary support is one of the most important factors determining its mechanical rigidity. In our opinion, multisupport mounting with a radial-symmetric arrangement of the supports is the most suitable one for a radial-symmetric (cyclic) loading scheme of the mirror frame. In particular, hub mounting of the mirror frame constitutes a radial-symmetric mounting system.

In fact, while it is radial-symmetric, a hub mounting does not constitute a multisupport mirror mounting. It is referred to as a cantilever system. While it is structurally the simplest system, a cantilever mounting has a considerable disadvantage: the basic load-carrying elements of the mirror frame, the radial elements, operate as cantilevers and are thus subject to considerable strain.

Figure 1 shows the schematic diagram of a cantilever mounting similar to that of the Australian radio telescope located near Sydney [1]. Strictly speaking, it could be referred to as a radial-symmetric (cyclic) system if the diameter of the hub were "equal to zero."

In practice, this means that not only the relative, but also the absolute, dimensions of the hub must be small if the hub is to be considered as absolutely rigid. It is seen in Fig. 1 that the mirror hub has two support pivots, so that the deflections (vertical displacements) at different points of its contour (in the plan) are unequal, i.e., the greater the distance from the support pivots, the larger the deflection. The deflections are at a maximum at two points on the diameter perpendicular to the line of the support pivots. However, the rigidity of the hub can be considerably increased by using a multisupport mounting, in particular, a four-point support, which is readily provided by means of a two-sector structure. Two rotation sectors with two support pivots provide four equally rigid supports for the hub, whereby its rigidity is increased by almost one order of magnitude.

In order to arrive at acceptable absolute dimensions of the hub for its assigned strain, we shall consider the deflection of the hub as a function of its dimensions.

Figure 2a shows the schematic diagram of the hub, constructed in the shape of a regular dioctahedron, which is bound by inside radial tie ribs. Figure 2b shows the rotation sectors with the horizontal axle and the support pivots. The diametric ribs, which are equally spaced with respect to the supports, are designated by the same figures; the near ones are marked by 1-1, and the distant ones are marked by 2-2. It is seen from Fig. 2b that the four support points on the rotation sectors are characterized by equal rigidity relative to the support pivots. This is the simplest way of transforming the two supports at the pivots into four equal supports for the hub. However, this does not secure complete radial symmetry of the support mounting (suspension) of the hub. A minimum of eight supports, one alternating with every two neighboring ribs, is necessary for complete radial symmetry.

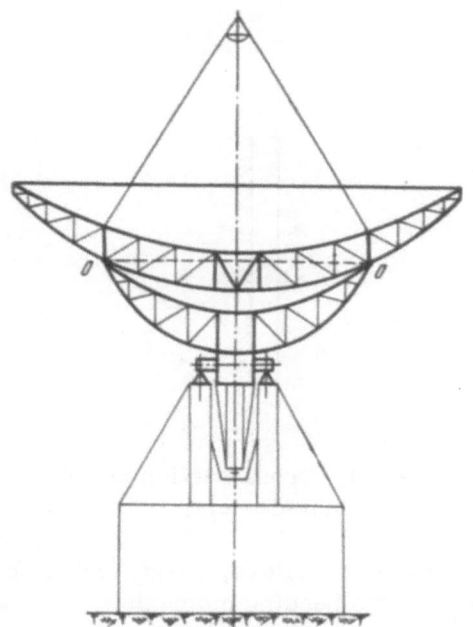

Fig. 1. Mirror with hub mounting (cantilever mounting).

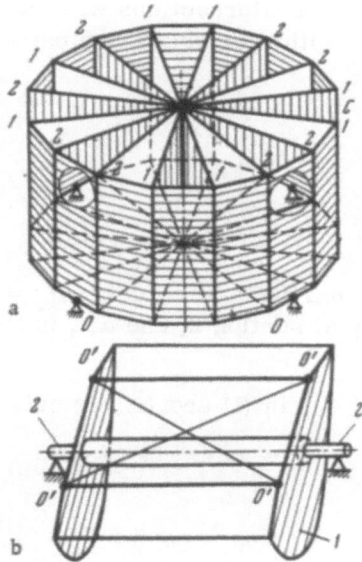

Fig. 2. Loading scheme of (a) the parabolic mirror hub, (b) the rotation sector unit with support pivots. 1) Rotation sectors; 2) support pivots; 1-1, 2-2) diametric elements (ribs); 0) hub support points; 0') critical hub support points on the rotation sectors.

In the case of eight-support radial-symmetric mounting of the hub, the problem of its intrinsic rigidity is eliminated, since it does not affect the distortion of the mirror.* However, since eight-support hub mounting requires a more complicated structure, we shall limit our considerations to the rigidity of a hub with four supports.

Let us consider the maximum deflections of a hub due to its own weight, with its axis in the vertical position (see Fig. 2a). The maximum deflections will be at the points C, positioned symmetrically between the supports.† Due to the structural and loading symmetry, the hub is twice statically indeterminate. It is sufficient to determine the stresses and strains in the elements forming one-eighth of the hub structure, since the stresses and strains repeat themselves in all the other corresponding elements.

Figure 3a shows the basic system for one-eighth of the hub. The support rods (ties) K replace the action of the discarded part of the hub. A small part of the wall, 0-0-Hi-Hi, has been removed and replaced by two hinges, Hi. The redundant unknown X_1 constitutes the shearing force of interaction between the two intersecting diametric ribs 1-1 and 2-2; the redundant unknown X_2 represents the bending moment exerted by the rejected part of the hub. It should be noted that the rotation of sections of the hub ring above the supports and at the points C (at midpoints between adjacent supports) is equal to zero due to symmetry. The bending moment X_2 in the ring remains constant because of symmetry [2]; the bending moments due to other loads are distributed throughout the ring as in a straight plane beam. Therefore, the system shown in Fig. 3b constitutes the basic system for the ring. Here, q (kgf/cm) is the uniformly distributed load due to the hub weight, P (kg) is the concentrated load due to the rib weight, and X_1 and X_2 are the redundant unknowns acting on the ring. Figure 3c,d

---

*Naturally, the hub must possess the minimum rigidity required of an ordinary metal structure.
† We are not interested in the deflection of the hub's central point on the geometric axis, although it could be larger than the deflection at the point C.

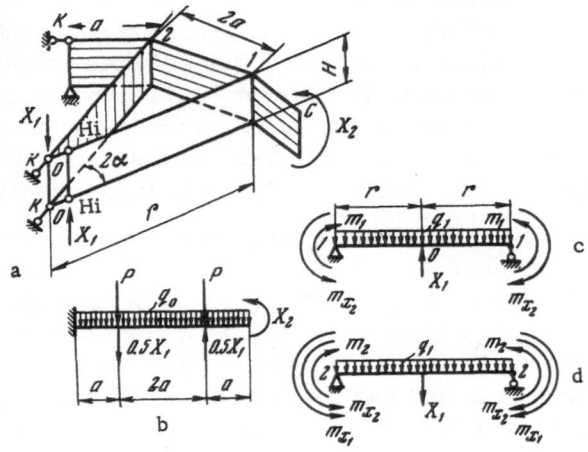

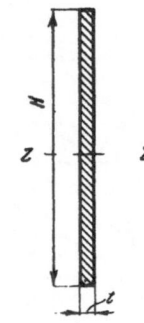

Fig. 3. Basic system for a part of the hub.

Fig. 4. Cross section of the hub element.

shows the basic systems for the diametric elements (ribs) of the hub. Here, q (kgf/cm) is the uniformly distributed load due to the rib weight, m_1 and m_2 are the bending moments in the radial elements 1 and 2, respectively, caused by the bending moments due to the loads q and P, m_{X_1} is the bending moment in the radial element 2 due to the action of X_1 on the ring beam, and m_{X_2} are the bending moments in the radial elements, caused by the action of X_2.

The bending moments m_1, m_2, m_{X_1}, and m_{X_2} arise in the radial elements due to the fact that the ring beam does not lie in the same plane and has breaks at intersections with the radial elements. * The above bending moments are determined by the following relationships:

$$m_1 = 2m_{k_1} \sin \alpha,$$
$$m_2 = 2m_{k_2} \sin \alpha; \tag{1}$$

$$m_{x_1} = 2aX_1 \sin \alpha,$$
$$m_{x_2} = 2X_2 \sin \alpha, \tag{2}$$

where $m_{k_1} = 0.5g_0 a^2$ is the bending moment due to the load in the ring at section 1, $m_{k_2} = 4.5g_0 a^2 + 2aP$ is the bending moment due to the load in the ring at section 2, and aX_1 is the bending moment in the ring due to X_1 at section 2.

By substituting the m_{k_1} and m_{k_2} values for $X_1 = 1$ and $X_2 = 1$ in (1) and (2), we obtain

$$m_1 = g_0 a^2 \sin \alpha,$$
$$m_2 = (9g_0 a^2 + 4aP) \sin \alpha,$$
$$m_{x_1} = 2a \sin \alpha, \tag{3}$$
$$m_{x_2} = 2 \sin \alpha.$$

For the sake of simplicity, we assume that the ring and the radial elements of the hub are simple plane beams with a rectangular cross section (Fig. 4). The moment of inertia of the

---

* The bending moments acting in the ring at the breaks of the ring beam (the line of intersection between the ring and a radial element) are decomposed into two components, one acting in the plane of the radial element and the other in the plane of the new direction of the ring element.

transverse cross section of such a beam relative to the Z–Z axis is given by

$$I = \frac{tH^3}{12}.$$ (4)

The cross section of the beams is designed with a view to decreasing their flexibility by concentrating much of the cross-sectional area in the top and the bottom parts of the beam (thus reinforcing its flanges) rather than distributing it uniformly along the height of the cross section, whereby the moment of inertia I is considerably increased. Therefore, by considering the cross section shown in Fig. 4, we drastically reduce the I value.

We shall express the quantities a, H, t, g_0, and P in terms of the hub radius r under the condition that the critical stress in the beam, which is considered as a plate, remains approximately constant with an increase in the dimensions a and H of the panel.

The critical stress for a compressed plate is given by [3]

$$\sigma_{cr} = k \frac{\pi^2 D}{H^2 t}.$$ (5)

Here, $D = Et^3/12(1-\mu^2)$ is the cylindrical rigidity of the plate, E (kgf/cm²) is the elasticity modulus of the first kind, and $\mu \approx 0.3$ is the Poisson coefficient (for steel). By substituting the D value in (5), we obtain

$$\sigma_{cr} = \frac{k\pi^2 E}{12(1-\mu^2)} \frac{t^2}{H^2} = k_1 \frac{t^2}{H^2}.$$ (6)

For σ_{cr} = const, the following condition must hold:

$$t/H = \text{const.}$$

Since, on the basis of the stringent rigidity requirements, the hub elements operate under low stresses, we shall use the value $\sigma_{cr} \approx 1200$ kgf/cm² (each face of the ring and the rib is divided into four panels by local ribs, so that a side ratio equal to a/H is maintained for each panel). In this case, the beam thickness can be expressed as t = 0.004r. Furthermore, we assume that H = 1.5r. For a dioctahedral hub, $\alpha = 11°15'$, $\sin\alpha \approx 0.195$, and $a = r\sin\alpha \approx 0.195r$. The cross-sectional area of a ring element is $F_0 = Ht = 0.006r^2$. The moment of inertia of the transverse cross section of a ring element is

$$I_0 = \frac{tH^3}{12} = \frac{0.004r(1.5r)^3}{12} = 112.50 \cdot 10^{-5} \, r^4.$$

The linear weight is

$$g_0 = F_0\gamma = 0.006 \, r^2\gamma.$$

For steel, the specific weight is $\gamma = 0.00785$ kgf/cm², and

$$g_0 = 4.72 \cdot 10^{-5} \, r^2.$$

For the concentrated load (rib weight),*

$$P = 0.5 \, rg_0 = 2.36 \cdot 10^{-5} \, r^3.$$

---

*We assume that the cross section of the radial elements — ribs — is one half as large as the cross section of ring beams.

The bending moments (3) are now written:

$$m_1 = 0.35 \cdot 10^{-6} \, r^4,$$
$$m_2 = 6.740 \cdot 10^{-6} \, r^4,$$
$$m_{x_1} = 0.076 \, r,$$
$$m_{x_2} = 0.39.$$

(7)

The redundant unknowns are determined by solving the system of equations

$$\delta_{11} X_1 + \delta_{12} X_2 + \Delta_{1p} = 0,$$
$$\delta_{21} X_1 + \delta_{22} X_2 + \Delta_{2p} = 0,$$

(8)

where the coefficients in front of the unknowns and the free terms are determined by the Maxwell — Mohr equations [4]:

$$\delta_{11} = \frac{1}{E} \int \frac{M_1^2}{I} \, dx + \frac{1}{G} \, \mu \int \frac{Q_1^2}{F} \, dx,$$

$$\delta_{22} = \frac{1}{E} \int \frac{M_2^2}{I} \, dx + \frac{1}{G} \, \mu \int \frac{Q_2^2}{F} \, dx,$$

$$\delta_{12} = \frac{1}{E} \int \frac{M_1 M_2}{I} \, dx + \frac{1}{G} \, \mu \int \frac{Q_1 Q_2}{F} \, dx,$$

$$\Delta_{1p} = \frac{1}{E} \int \frac{M_1 M_p}{I} \, dx + \frac{1}{G} \, \mu \int \frac{Q_1 Q_p}{F} \, dx,$$

$$\Delta_{2p} = \frac{1}{E} \int \frac{M_2 M_p}{I} \, dx + \frac{1}{G} \, \mu \int \frac{Q_2 Q_p}{F} \, dx,$$

(9)

where M_1, M_2, Q_1, and Q_2 are the bending moments and the shearing forces in the hub elements, respectively, caused by unit values of the unknowns, M_p and Q_p are the bending moment and the shearing force in the hub elements due to the load, respectively, E and G are the elasticity moduli of the first and the second kind, respectively, F is the cross-sectional area of the element (beam), and μ is the cross-section factor.

After integrating (9) by using Vereshchagin's method (multiplication of curves) [4], we obtain the following values for the coefficients and free terms:

$$E\delta_{11} = 887.5 \, r^{-1},$$
$$E\delta_{22} = 1235 \, r^{-3},$$
$$E\delta_{12} = 120.5 \, r^{-2},$$
$$E\Delta_{1p} = 0.00923 \, r^2,$$
$$E\Delta_{2p} = -0.035 \, r.$$

(10)

By solving the system of equations (8) and considering (10), we obtain

$$X_1 = -14.46 \cdot 10^{-6} \, r^3,$$
$$X_2 = + 30.20 \cdot 10^{-6} \, r^4.$$

(11)

The maximum deflection at the point C (see Fig. 3a), found by means of the expression [4]

$$\delta_C = \frac{1}{EI} \int M_C M_\Sigma dx + \frac{\mu}{GF} \int Q_C Q_\Sigma dx,$$

(12)

where M_C and Q_C are the bending moment and the shearing force in the hub elements due to the unit force applied at the point C, M_Σ and Q_Σ are the calculated bending moment and shearing

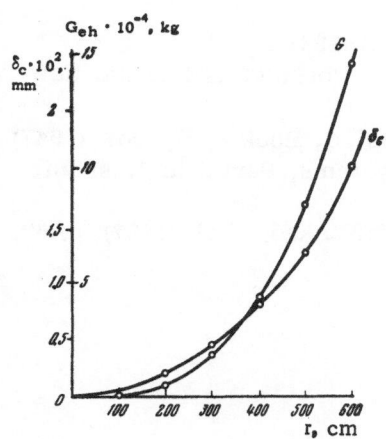

Fig. 5. Maximum deflection δ_C at the hub rim and the hub weight G_{eh} as functions of the radius r.

force (due to the combined action of the load, X_1, and X_2), is given by

$$\delta_C = 50 \cdot 10^{-10}\, r^2. \tag{13}$$

It is obvious from (13) that the maximum deflection of the hub in between adjacent supports is proportional to the square of its radius. The weight of the hub increases as the third power of its radius and is determined by the relationship

$$G_{eh} = 67.4 \cdot 10^{-5}\, r^3. \tag{14}$$

In expressions (13) and (14), r is in centimeters, while δ_C and G_{eh} are given in centimeters and kilograms-force, respectively.

Thus, in spite of the fact that the moment of inertia of the hub elements increases as the fourth power of the hub radius:

$$I_0 = 112.50 \cdot 10^{-5}\, r^4, \tag{15}$$

the deflections of the hub rapidly increase with its dimensions, as is shown by (13). Figure 5 shows δ_C and G_{eh} as functions of the hub radius r. It is seen that the maximum deflection at the rim of the hub due solely to its own weight is equal to 0.0045 mm for r = 300 cm (D = 600 cm).

We shall assume that the hub diameter amounts to 10% of the mirror diameter, while the mirror weight is determined (roughly) by the relationship [5]:

$$G_{mir} = 28 D^{2.5}. \tag{16}$$

Then, the concentrated load on a ring element of the hub (see Fig. 3) is given by

$$P_1 = P + \frac{1}{16} G_{mir} = P + \frac{28}{16} D^{2.5}, \tag{17}$$

while the maximum deflection at its rim is

$$\delta_C \approx \left(\frac{16 P_1}{G_{eh}} \delta_C \right). \tag{18}$$

For r = 300 cm, P = $0.5 g_0 r$ = 637 kg, G_{eh} = 18,140 kg (Fig. 5), D_{mir} = 10 · 2r = 6000 cm = 60 m, G_{mir} = 28 · $D^{2.5}$ = 28 · $60^{2.5}$ = 780,000 kgf. By substituting these values in (17) and (18), we obtain

$$(\delta_C)_1 = \frac{16}{18,140} \delta \left(637 + \frac{1}{16} \cdot 780,000 \right) \approx 43.4\, \delta_1.$$

For δ_1 = 0.0045 mm, $(\delta_C)_1 \approx$ 0.2 mm.

If the mirror hub had two supports, the maximum deflection at its rim (in between the supports) would be approximately eight times as large, i.e., $\delta_1' \approx 8 \cdot 0.2 = 1.6$ mm.

We thus reach the conclusion that, for large parabolic mirrors designed for operation at wavelengths of 1-2 cm, the distortion of the mirror shape in the central (and the most important) part due only to nonuniform deformations of the hub assumes considerable proportions if the hub has large absolute dimensions (in excess of 5 m) and is mounted on two supports.

Since the size of the hub considerably affects the rigidity of the mirror mounting, a four-support mounting is highly desirable because it would make it possible to increase both the dimensions of the hub and its rigidity in comparison with a two-support mounting.

LITERATURE CITED

1.   E. G. Bowen and H. C. Minnett, Proc. IRE Australia, 24, No. 2 (1963).
2.   P. D. Kalachev, in: Radio Telescopes (D. V. Skobel'tsyn, ed.), Consultants Bureau, New York (1966), p. 143.
3.   Encyclopedic Mechanical Engineering Manual [in Russian], Vol. 1, Book 2, Moscow (1947).
4.   I. M. Rabinovich, Textbook of Structural Mechanics of Rod Systems, Part 2 [in Russian], Gosstroiizdat (1940).
5.   B. A. Garf, Utilization of Solar Energy, Vol. 1 [in Russian], Izd. Akad. Nauk (1957), p. 62.

DESIGN ELASTIC DEFORMATIONS OF A 7.5-METER PARABOLIC MIRROR

P. D. Kalachev, V. P. Nazarov,
V. Ya. Chashnikov, and A. A. Parshchikov

Introduction

The interest of specialists in radio telescopes with fully steerable parabolic mirror antennas has recently increased markedly.* This is explained by the fact that these antennas have certain advantages, which we will not dwell upon in this article. One of the most important problems which must be solved when creating large fully steerable parabolic mirrors with a high resolution is to provide a high mechanical rigidity of the mirror at any position in space. The main difficulty in meeting this provision is that the mirror changes its position in space [1, 2].

Parabolic mirrors of radio telescopes are calculated mainly for two cases of loading: a static calculation of the mirror for rigidity in the case of the effect of service loads and a static calculation of the mirror for strength in the case of hurricane-force wind. A dynamic calculation of the mirror for the effect of inertial forces is usually not performed, since the velocities and accelerations during rotations of the mirror are insignificant.

The service loads on the mirror are the loads from the dead weight of the structure and wind loads. For mirrors with a high rigidity and hence with a large dead weight, the weight loads are on the order of 100–200 kg/m² of area of mirror aperture. The wind loads at a service wind velocity taken to be 8-12 m/sec are of the order of 8-16 kgf/m². When calculating wind loads for a hurricane-force wind the speed of the latter is taken to be 40-50 m/sec.

The problem in the first case of calculation consists in determination of elastic deformations of the mirror and in selection of the cross sections of the force-bearing members of the mirror's construction (for a given force-bearing scheme) such that they provide minimum magnitudes of elastic deformations. The cross sections of the force-bearing members selected on the basis of this condition generally provide the strength necessary for the second case of calculation, i.e., under loads from hurricane-force wind.

A simple analysis of a parabolic mirror as a structural member and also the experience of designing telescopes with fully steerable parabolic mirrors show that it is not possible by simply increasing the cross sections of the force-bearing members of the mirror frame to achieve a substantial increase of its rigidity, although within certain limits an increase in the sections of the main supporting members (radial members) gives a noticeable gain [3]. A radical solution of this problem is to decrease the "design" lengths of the force-bearing elements,

---

*This was evidenced at the conferences at London and Munich in 1966.

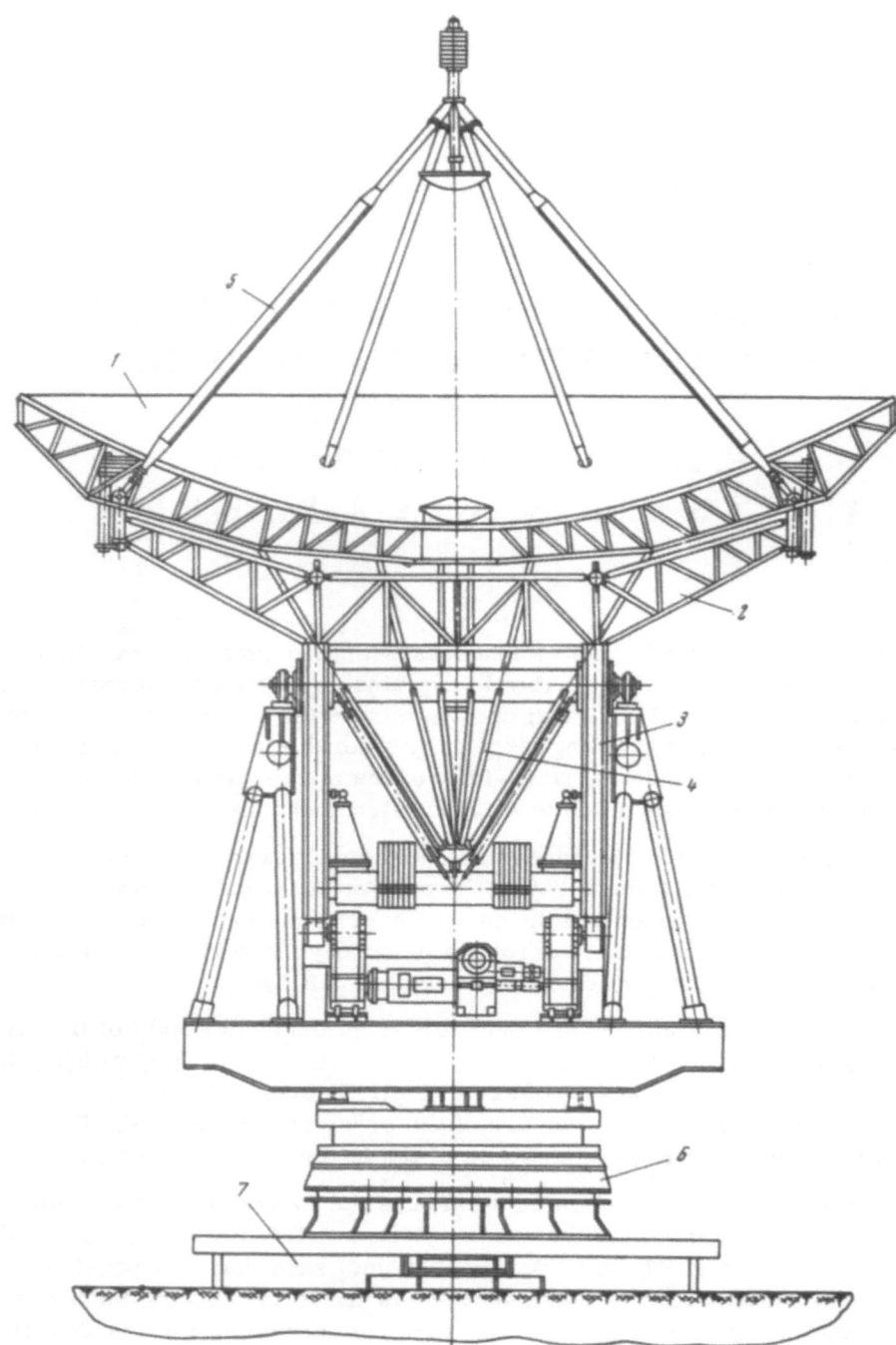

Fig. 1. General view of a radio telescope with a 7.5-m parabolic mirror on a multiple-support structure. 1) Parabolic mirror; 2) intermediate eight-point support
structure; 3) rotating sectors assembly; 4) nine-bar pyramid; 5) eight-bar pyramid
supporting the exciter; 6) support-rotating unit; 7) foundation.

i.e., decrease the cantilever and span portions of the mirror. This can be achieved by increasing the number of mirror supports.

This article presents a static calculation of a model of a 7.5-m diameter, fully-steerable
parabolic mirror on a multiple-support structure. The results of this calculation show that

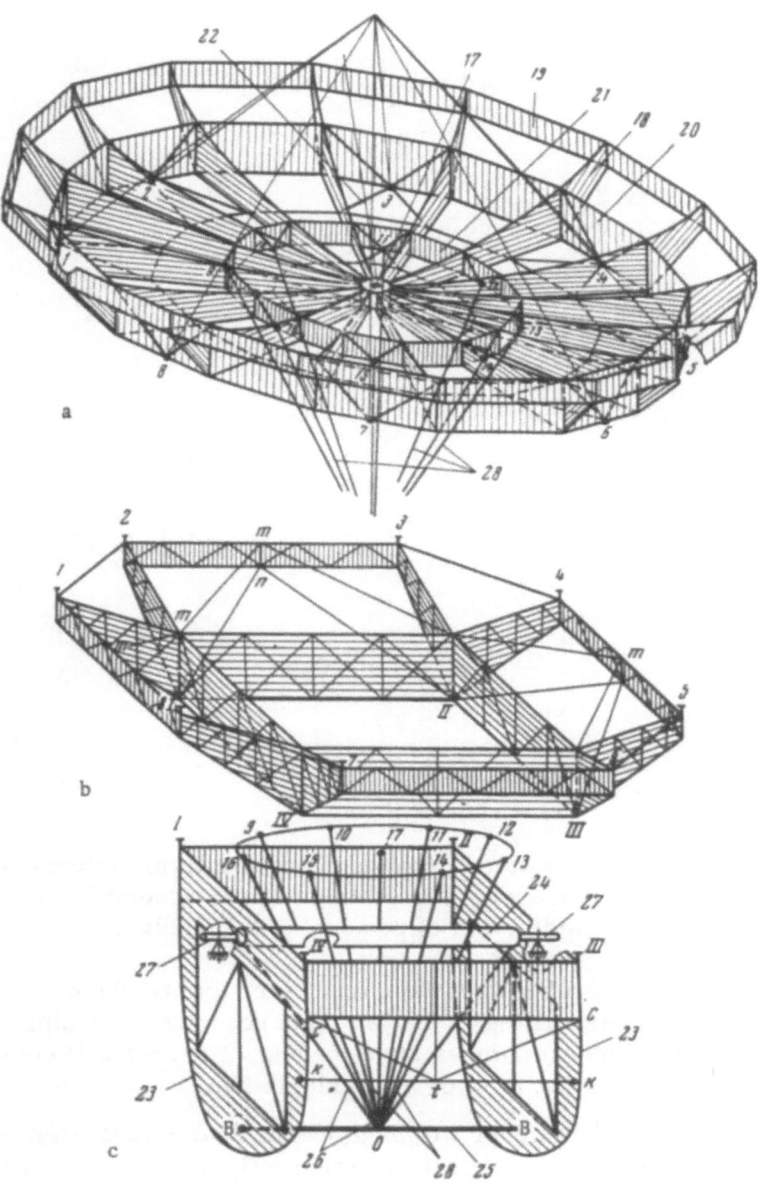

Fig. 2. Force diagram of mirror and its supporting structure in an exploded form. a) Mirror; b) intermediate eight-point support structure; c) rotating sectors assembly. I, II, III, IV) support points of intermediate structure on rotating sectors; 1, 2, . . . , 8) main support points of mirror formed by intermediate structure; 9, 10, . . . , 17) auxiliary support points of mirror formed by nine-bar pyramid with central bar; 18) radial force-bearing members; 19, 20, 21) arc (annular) members; 22) knee braces in lower panels of frame; 23) rotating sectors; 24) horizontal axle; 25) counterbalance beam; 26) knee braces; 27) support trunnions; 28) nine-bar pyramid.

the rigidity of the mirror on a 17-point support structure is almost an order of magnitude higher than the rigidity of a mirror on the conventional two-point support structure [1]. A general view of the 7.5-m model of the fully steerable parabolic antenna is shown in Fig. 1.

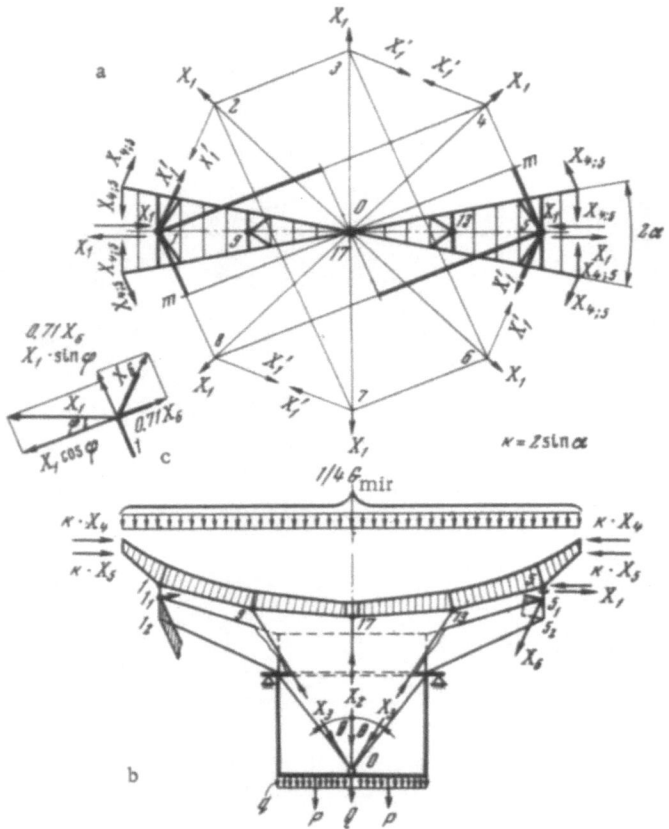

Fig. 3. Basic system of isolated part of mirror, intermediate support structure, and rotating sectors assembly. a) Plan view; b) side view; c) removal of assembly 1.

In the first case (static calculation of the mirror for rigidity) the calculation is performed according to two schemes corresponding to the two main schemes of loading of the mirror: symmetric loading of the mirror and skew-symmetric loading. Here we will consider the symmetric scheme of loading the mirror for the first case of calculation.

Since we are interested in a mirror with high mechanical rigidity, the main design loads will be those from the structure's own weight. Consequently, symmetric loading corresponds to the horizontal position of the aperture plane of the mirror. Figure 2 shows the force diagram of the mirror and its supporting structure in an exploded isometric projection. The inside diagonal braces m-m and I-n of the intermediate structure are not affected by symmetric loading and therefore they are depicted by thin lines. In the force diagram of the mirror (Fig. 2a) the knee braces in the lower panel, which are not affected by symmetric loads, are not shown. In order not to obscure the drawing, members k-k and c-t, which increase the flexural and torsional rigidity of the sectors assembly, are depicted arbitrarily as thin lines.

As a consequence of the symmetry of the structure and load, the forces in all similar force-bearing members, i.e., members arranged radially symmetric relative to the center of the mirror, will be the same. Taking into consideration the conditions of symmetry, as the basic system for the mirror we will take one-quarter of it bounded by two adjacent, mutually intersecting diametral force-bearing members (trusses) and by the arc members and knee braces, which come up to the support assemblies, which enter this part of the mirror. As the

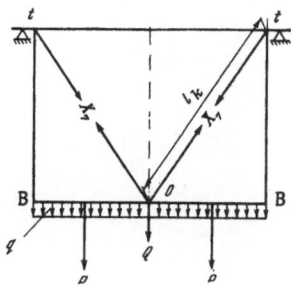

Fig. 4. Basic system for the sectors assembly (l_k is the length of the knee braces).

basic system for the intermediate structure we take also one-quarter of it, including the main truss (for example, 8-5) and the end semitrusses 5-m and 8-m. The main system for the rotating sectors assembly includes: props N-T (needle beams), counterbalance beam B-B, and horizontal axle 27-27. The indicated systems are depicted arbitrarily superimposed in Fig. 3.

The arbitrariness here is that, in reality, there is no single force-bearing member located along the diagonal of the intermediate structure, but by virtue of the symmetry all force-bearing members of the mirror and of the intermediate structure work alike. Therefore we can consider that the two support points 1 and 5 are formed by one main truss of the intermediate structure (for example, 8-5 together with its strengthening members: struts 6-5 and 7-8 and end semitrusses 5-m and 8-m), although, as Fig. 3 shows, the scheme includes semitrusses 1-4 and 5-8, 1-m and 5-m.

For the redundant quantities we take: X_1, the thrust force between the mirror frame and the intermediate structure; X_2 and X_3, the forces acting in the bars of the nine-bar pyramid; X_4 and X_5, the forces acting in the outside arc (annular) member in the top and bottom chords, respectively; X_6, the force in the struts of the intermediate structure (see Fig. 3); X_7, the force in the knee braces of the joint of the sectors (Fig. 4).

We note that the system: mirror, intermediate structure, and rotating sectors assembly, is naturally, as it were, separated into three independent units. Thus, the forces in the members of the rotating sectors assembly from the effect of $X_7 = 1$ (Fig. 4) are determined independently of the mirror and intermediate structure, and the forces in the members of the intermediate structure from the effect of $X_6 = 1$ are determined independently of the mirror and sectors assembly (see Figs. 2 and 3).

The system of canonical equations relating the redundant quantities $X_1, \ldots, X_7$ has the form

$$
\begin{aligned}
&\delta_{11}X_1 + \delta_{12}X_2 + \delta_{13}X_3 + \ldots + \delta_{17}X_7 + \Delta_{1p} = 0, \\
&\delta_{21}X_1 + \delta_{22}X_2 + \delta_{23}X_3 + \ldots + \delta_{27}X_7 + \Delta_{2p} = 0, \\
&\delta_{31}X_1 + \delta_{32}X_2 + \delta_{33}X_3 + \ldots + \delta_{37}X_7 + \Delta_{3p} = 0, \\
&\cdots\cdots\cdots\cdots\cdots\cdots\cdots\cdots\cdots\cdots \\
&\delta_{71}X_1 + \delta_{72}X_2 + \delta_{73}X_3 + \ldots + \delta_{77}X_7 + \Delta_{7p} = 0.
\end{aligned}
\tag{1}
$$

The coefficients and free terms of Eqs. (1) are determined by the known Maxwell — Mohr formulas:

$$
\begin{aligned}
\delta_{ii} &= \Sigma \overline{X}_i^2 \frac{l}{EF} + \int \overline{M}_i^2 \frac{dx}{EI} + \mu \int \overline{Q}_i^2 \frac{dx}{GE_{\mathrm{bar}}}, \\
\delta_{ik} &= \Sigma \overline{X}_i \overline{X}_k \frac{l}{EF} + \int \overline{M}_i \overline{M}_k \frac{dx}{EI} + \mu \int \overline{Q}_i \overline{Q}_k \frac{dx}{GF_{\mathrm{bar}}}, \\
\Delta_{ip} &= \Sigma \overline{X}_i N_p \frac{l}{EF} + \int \overline{M}_i M_p \frac{dx}{EI} + \mu \int \overline{Q}_i Q_p \frac{dx}{GF_{\mathrm{bar}}},
\end{aligned}
\tag{2}
$$

where the first terms are the displacements due to deformation of the bars and the second and third terms are displacements due to bending and shear in the beams, respectively.

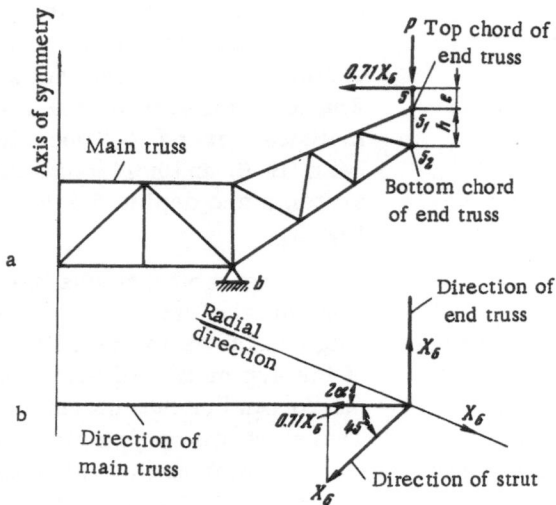

Fig. 5. Main semitruss of intermediate support structure. a) Side view; b) plan view; $P = \frac{1}{8}(G_{mir} + G_{exc})$.

In our case the coefficients and free terms of Eqs. (1) are determined by the following relations:

$$\delta_{11} = \left(\sum \overline{X}_1^2 \frac{l}{EF}\right)_{\frac{1}{8}\,mir} + \left(\sum \overline{X}_1^2 \frac{l}{EF}\right)_{\frac{1}{8}\,i.s} + \int \overline{M}_1^2 \frac{dx}{ET_{s.b}} + \mu \int \overline{Q}_1^2 \frac{dx}{GF_{s.b}},$$

$$\delta_{12} = \left(\sum \overline{X}_1 \overline{X}_2 \frac{l}{EF}\right)_{\frac{1}{8}\,mir} + \left(\sum \overline{X}_1 \overline{X}_2 \frac{l}{EF}\right)_{\frac{1}{8}\,i.s},$$

$$\delta_{13} = \left(\sum \overline{X}_1 \overline{X}_3 \frac{l}{EF}\right)_{\frac{1}{8}\,mir} + \left(\sum \overline{X}_{13} \overline{X} \frac{l}{EF}\right)_{\frac{1}{8}\,i.s},$$

$$\delta_{14} = \left(\sum \overline{X}_1 \overline{X}_4 \frac{l}{EF}\right)_{\frac{1}{8}\,mir},$$

$$\delta_{15} = \left(\sum \overline{X}_1 \overline{X}_5 \frac{l}{EF}\right)_{\frac{1}{8}\,mir},$$

$$\delta_{16} = \left(\sum \overline{X}_1 \overline{X}_6 \frac{l}{EF}\right)_{\frac{1}{8}\,i.s},$$

$$\delta_{17} = 0, \qquad \delta_{21} = \delta_{12},$$

$$\delta_{22} = \left(\sum \overline{X}_2^2 \frac{l}{EF}\right)_{\frac{1}{8}\,mir} + \left(\sum \overline{X}_2^2 \frac{l}{EF}\right)_{\frac{1}{8}\,i.s} + \left(\sum \overline{X}_2^2 \frac{l_{prop}}{EF'_{prop}}\right)_{\frac{1}{2}\,sec.a} +$$

$$+ \overline{X}_2^2 \frac{l'_{bar\,p}}{EF_{bar\,p}} + \int \overline{M}_2^2 \frac{dx}{EI'_{c.b}} + \mu \int \overline{Q}_2^2 \frac{dx}{GF'_{c.b}},$$

$$\delta_{23} = \left(\sum \overline{X}_2 \overline{X}_3 \frac{l}{EF}\right)_{\frac{1}{8}\,mir} + \left(\sum \overline{X}_2 \overline{X}_3 \frac{l}{EF}\right)_{\frac{1}{8}\,i.s} + \left(\sum \overline{X}_2 \overline{X}_3 \frac{l_{prop}}{EF'_{prop}}\right)_{\frac{1}{2}\,sec.a} +$$

$$+ \int \overline{M}_2 \overline{M}_3 \frac{dx}{EI_{c.b}} + \mu \int \overline{Q}_2 \overline{Q}_3 \frac{dx}{GF_{c.b}},$$

$$\delta_{24} = \left(\sum \overline{X}_2 \overline{X}_4 \frac{l}{EF}\right)_{\frac{1}{8}\,mir},$$

$$\delta_{25} = \left(\sum \overline{X}_2 \overline{X}_5 \frac{l}{EF}\right)_{\frac{1}{8}\,mir},$$

$$\delta_{26} = \left(\sum \overline{X}_2 \overline{X}_6 \frac{l}{EF}\right)_{\frac{1}{8}\,i.s},$$

(3)

$$\delta_{27} = \left(\sum \overline{X}_2 \overline{X}_7 \frac{l_{\text{prop}}}{EF'_{\text{prop}}}\right)_{\frac{1}{2}\text{sec.a}} + \int \overline{M}_2 \overline{M}_7 \frac{dx}{EI'_{\text{c.b}}} + \mu \int \overline{Q}_2 \overline{Q}_7 \frac{dx}{GF'_{\text{c.b}}},$$

$$\delta_{31} = \delta_{13}, \quad \delta_{32} = \delta_{23},$$

$$\delta_{33} = \left(\sum \overline{X}_3^2 \frac{l}{EF}\right)_{\frac{1}{8}\text{mir}} + \left(\sum \overline{X}_3^2 \frac{l}{EF}\right)_{\frac{1}{8}\text{i.s}} + \left(\sum \overline{X}_3^2 \frac{l_{\text{prop}}}{EF'_{\text{prop}}}\right)_{\frac{1}{8}\text{sec.a}} +$$
$$+ \int \overline{M}_3^2 \frac{dx}{EI'_{\text{c.b}}} + \mu \int \overline{Q}_3^2 \frac{dx}{GF'_{\text{c.b}}} + \overline{X}_3^2 \frac{l_{\text{bar p}}}{EF_{\text{bar p}}},$$

$$\delta_{34} = \left(\sum \overline{X}_3 \overline{X}_4 \frac{l}{EF}\right)_{\frac{1}{8}\text{mir}},$$

$$\delta_{35} = \left(\sum \overline{X}_3 \overline{X}_5 \frac{l}{EF}\right)_{\frac{1}{8}\text{mir}},$$

$$\delta_{36} = \left(\sum \overline{X}_3 \overline{X}_6 \frac{l}{EF}\right)_{\frac{1}{8}\text{i.s}},$$

$$\delta_{37} = \left(\sum \overline{X}_3 \overline{X}_7 \frac{l_{\text{prop}}}{EF'_{\text{prop}}}\right)_{\frac{1}{2}\text{sec.a}} + \int \overline{M}_3 \overline{M}_7 \frac{dx}{EI'_{\text{c.b}}} + \mu \int \overline{Q}_3 \overline{Q}_7 \frac{dx}{GF'_{\text{c.b}}},$$

$$\delta_{41} = \delta_{14}, \quad \delta_{42} = \delta_{24}, \quad \delta_{43} = \delta_{34},$$

$$\delta_{44} = \left(\sum \overline{X}_4^2 \frac{l}{EF}\right)_{\frac{1}{8}\text{mir}} + \overline{X}_4^2 \frac{l_{\text{low fr}}}{F'_{\text{low fr}}},$$

$$\delta_{45} = \left(\sum \overline{X}_4 \overline{X}_5 \frac{l}{EF}\right)_{\frac{1}{8}\text{mir}},$$

$$\delta_{46} = 0, \quad \delta_{47} = 0,$$
$$\delta_{51} = \delta_{15}, \quad \delta_{52} = \delta_{25}, \quad \delta_{53} = \delta_{35}, \quad \delta_{54} = \delta_{45},$$

$$\delta_{55} = \left(\sum \overline{X}_5^2 \frac{l}{EF}\right)_{\frac{1}{8}\text{mir}} + \overline{X}_5^2 \frac{l_{\text{low fr}}}{F'_{\text{low fr}}},$$

$$\delta_{56} = 0, \quad \delta_{57} = 0,$$
$$\delta_{61} = \delta_{16}, \quad \delta_{62} = \delta_{26}, \quad \delta_{63} = \delta_{36}, \quad \delta_{64} = \delta_{46}, \quad \delta_{65} = \delta_{56},$$

$$\delta_{66} = \left(\sum \overline{X}_6^2 \frac{l}{EF}\right)_{\frac{1}{8}\text{i.s}} + \overline{X}_6^2 \frac{l_p}{EF_p},$$

$$\delta_{71} = \delta_{17}, \quad \delta_{72} = \delta_{27}, \quad \delta_{73} = \delta_{37}, \quad \delta_{74} = \delta_{47},$$
$$\delta_{75} = \delta_{57}, \quad \delta_{76} = \delta_{67},$$

$$\delta_{77} = \left(\sum \overline{X}_7^2 \frac{l_{\text{prop}}}{EF'_{\text{prop}}}\right)_{\frac{1}{2}\text{sec.a}} + \overline{X}_7^2 \frac{l_{\text{k.b}}}{EF'_{\text{k.b}}} + \int \overline{M}_7^2 \frac{dx}{EI'_{\text{c.b}}} + \mu \int \overline{Q}_7^2 \frac{dx}{GF'_{\text{c.b}}},$$

$$\Delta_{1p} = \left(\sum \overline{X}_1 N_p \frac{l}{EF}\right)_{\frac{1}{8}\text{mir}} \left(\sum \overline{X}_1 N_p \frac{l}{EF}\right)_{\frac{1}{3}\text{i.s}} + \int \overline{M}_1 M_p \frac{dx}{EI_{\text{s.b}}} + \mu \int \overline{Q}_1 Q_p \frac{dx}{GF_{\text{bar sup}}},$$

$$\Delta_{2p} = \left(\sum \overline{X}_2 N_p \frac{l}{EF}\right)_{\frac{1}{8}\text{mir}} + \left(\sum \overline{X}_2 N_p \frac{l}{EI}\right)_{\frac{1}{8}\text{i.s}} +$$
$$+ \left(\sum \overline{X}_2 N_p \frac{l_{\text{prop}}}{EF'_{\text{prop}}}\right)_{\frac{1}{2}\text{sec.a}} + \int \overline{M}_2 M_p \frac{dx}{FI_{\text{c.b}}} + \mu \int \overline{Q}_2 Q_p \frac{dx}{GF'_{\text{c.b}}},$$

$$\Delta_{3p} = \left(\sum \overline{X}_3 N_p \frac{l}{EF}\right)_{\frac{1}{8}\text{mir}} + \left(\sum \overline{X}_3 N_p \frac{l}{EF}\right)_{\frac{1}{8}\text{i.s}} +$$
$$+ \left(\sum \overline{X}_3 N_p \frac{l_{\text{prop}}}{EF'_{\text{prop}}}\right)_{\frac{1}{2}\text{sec.a}} + \int \overline{M}_3 M_p \frac{dx}{EI_{\text{c b}}} + \mu \int \overline{Q}_3 Q_p \frac{dx}{GF'_{\text{c.b}}},$$

$$\Delta_{4p} = \left(\sum \overline{X}_4 N_p \frac{l}{EF}\right)_{\frac{1}{8}\text{mir}}, \quad \Delta_{5p} = \left(\sum \overline{X}_5 N_p \frac{l}{EF}\right)_{\frac{1}{8}\text{mir}},$$

$$\Delta_{6p} = \left(\sum \overline{X}_6 N_p \frac{l}{EF}\right)_{\frac{1}{8}\text{i.s}};$$

$$\Delta_{7p} = \left(\sum \overline{X}_7 N_p \frac{l_{\text{prop}}}{EF'_{\text{prop}}}\right)_{\frac{1}{2}\text{sec. a}} + \int \overline{M}_7 M_p \frac{dx}{EI'_{\text{c.b}}} + \mu \int \overline{Q}_7 Q_p \frac{dx}{GF'_{\text{c.b}}}.$$

$$\left.\right\} \quad (3)$$

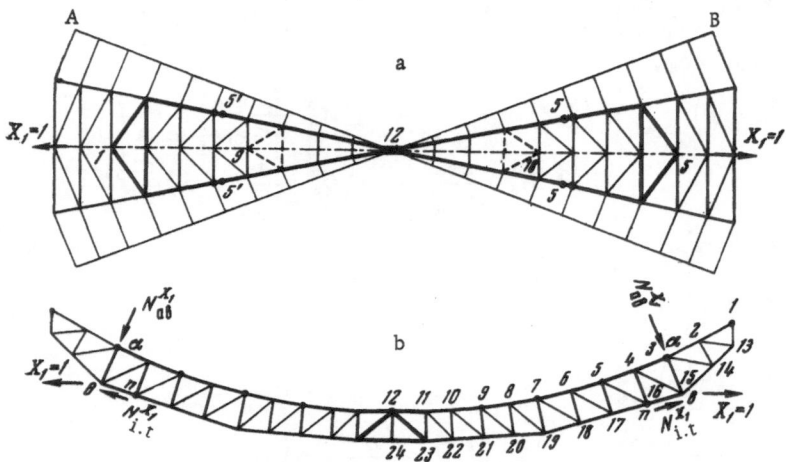

Fig. 6. Isolated portion of mirror with two adjacent, mutually intersecting diametral trusses. a) Plan view; b) side view.

Here $\overline{X}_1, \overline{X}_2, \ldots, \overline{X}_7$ are the forces in the bars of the system from the effect of unit values of
the unknowns $X_1, X_2, \ldots, X_7$; N_p are the forces in the same bars from the effect of the load
(dead weight of the structure); $\overline{M}_{1,2,\ldots,7}$ and $\overline{Q}_{1,2,\ldots,7}$ are the bending moments and transverse
forces arising in the beams of the system from the effect of unit values of the unknowns X_1, X_2,
$\ldots, X_7$; M_p and Q_p are the bending moments and transverse forces arising in the same beams
of the system from the effect of the load (dead weight of the structure); l and F are the lengths
and cross-sectional areas of the bars; E and G are the moduli of elasticity and shear of the material of the structure; $I'_{s.b.}$ and $F'_{s.b.}$ are the moment of inertia and cross-sectional area of
the support beam (prop);

$$l'_{c.b} = \frac{1}{4} I_{c.b}, \quad F'_{c.b} = \frac{1}{4} F_{c.b}, \quad \text{where } l_{c.b} \text{ and } F_{c.b}$$

are the moment of inertia and the cross-sectional area of the counterbalance beam (here we
take into account only $\frac{1}{4}$ of the beam section corresponding to $\frac{1}{4}$ of the mirror and intermediate
structure entering the calculation). Integration is carried out on half of the length of the counterbalance beam so that henceforth the calculation includes only half of the $\frac{1}{4}$ of the mirror and
intermediate structure ($\frac{1}{8}$ mir and $\frac{1}{8}$ i.s).

Terms of the form $\left(\sum \overline{X}_i^2 \dfrac{l}{EF} \right)_{\frac{1}{8} \text{ mir}}$, $\left(\sum \overline{X}_i \overline{X}_k \dfrac{l}{EF} \right)_{\frac{1}{8} \text{ mir}}$ and $\left(\sum \overline{X}_i N_p \dfrac{l}{EF} \right)_{\frac{1}{8} \text{ mir}}$ represent

displacements due to deformation of the bar members forming a part of the mirror frame, namely, two adjacent radial trusses and the arc trusses and knee braces abutting them in the lower

panels. Terms of the form $\left(\sum \overline{X}_i^2 \dfrac{l}{EF} \right)_{\frac{1}{8} \text{ i.s}}$, $\left(\sum \overline{X}_i \overline{X}_k \dfrac{l}{EF} \right)_{\frac{1}{8} \text{ i.s}}$ and $\left(\sum \overline{X}_i N_p \dfrac{l}{EF} \right)_{\frac{1}{8} \text{ i.s}}$ repre-

sent displacements due to deformation of the bar members forming $\frac{1}{8}$ of the intermediate eight-
point support structure, namely, the main semitruss, end semitruss, and half of the length of
the diagonal brace (1-2 or 3-4 in Figs. 2 and 3).*

The bar members of the rotating sector assembly are members lying in a vertical plane
passing through the support trunnions (pipes 27-27) serving as the horizontal axis of rotation,

---

*Knee braces 1-n and m-m (Fig. 2) are not affected by symmetric loads.

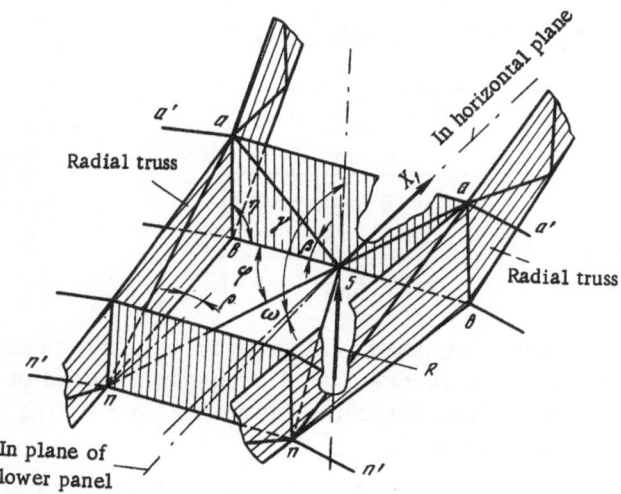

Fig. 7. Diagram of one of the main support assemblies
of the mirror.

props (needle beams) T-B, and counterbalance beam B-B. The force diagram of the rotating
sectors assembly in the presence of a symmetric load is shown in Fig. 4.

The loads on the rotating sectors assembly, i.e., on the bar members T-T, B-T, 0-t, and
on the counterbalance beam B-B, are: q, the counterbalance beam's own weight, which is a
uniformly distributed load; P, the load of the counterbalance weights p; and Q, the weight of the
nine-rod pyramid projected through its vertex q. We repeat that, for the design length of bars
T-T, B-T, and the counterbalance beam B-B, we take half of their geometric length, and for the
design cross-sectional area we take $\frac{1}{4}$ of the cross-sectional area of the corresponding member of the part of the mirror and intermediate structure being calculated.

The isolated part of the mirror (see Figs. 3 and 6) — two adjacent, mutually intersecting
diametral trusses with arc members and knee braces abutting them in the lower panels — is a
space bar system.

Figure 7 shows one support assembly of the mirror which is tied by an arc support truss
aa-bb and knee braces 5-n with radial trusses into a space bar system.

Decomposing the vertical support reaction R into components acting on bars 5-a, 5-n,
a-a, and b-b, we change to plane radial trusses. Thus we begin with force $X_1 = 1$.

The forces in bars 5-a, 5-n, a-a, and b-b from R and $X_1 = 1$ are determined by the following relations:

$$
\left.
\begin{aligned}
N^R_{5-a} &= \frac{R}{2\sin\beta\,(\cos\gamma\tan\omega + \sin\gamma)}, \\[4pt]
N^R_{a-a} &= \frac{R}{2}\,\frac{\cot\eta + \cot\beta}{\cos\gamma\tan\omega + \sin\gamma}, \\[4pt]
N^R_{a-b} &= \frac{R}{2\sin\eta\,(\cos\gamma\tan\omega + \sin\gamma)}, \\[8pt]
N^R_{5-n} &= \frac{R}{2\sin\varphi\,(\cos\omega\tan\gamma + \sin\omega)}, \\[4pt]
N^R_{n-n} &= \frac{R\,(\tan\rho + \cot\varphi)}{2\,(\cos\omega\tan\gamma + \sin\omega)},
\end{aligned}
\right\}
\qquad (4)
$$

$$N_{i.t}^{R} = \frac{R}{2\cos \rho \,(\cos \omega \tan \gamma + \sin \omega)},$$

$$N_{5-a}^{X_1} = \frac{X_1}{2\sin \beta \left(\dfrac{\sin \gamma}{\tan \omega} + \cos \gamma\right)},$$

$$N_{a-a}^{X_1} = \frac{X_1}{2\sin \eta \left(\dfrac{\sin \gamma}{\tan \omega} + \cos \gamma\right)},$$

$$N_{5-n}^{X_1} = \frac{X_1}{2\sin \varphi \left(\dfrac{\sin \omega}{\tan \gamma} + \cos \omega\right)}.$$

$$\text{(4)}$$

Similar relations are obtained for determining the forces in the corresponding bars of assembly 13 (or 9 in Figs. 2 and 3) from forces $X_3 = 1$ acting in the bars of the nine-bar pyramid 0–13 with consideration of angle θ. The diagram of forces $X_1 = 1$ acting on a system of two adjacent diametral trusses is shown in Fig. 6.

Formulas (4) are derived from the condition that the effect of the bars of the annular (arc) members a–a' and n–n' is taken into account by doubling the cross-sectional area of rods a–a and n–n. The numerical values of the coefficients and free terms calculated by formulas (3) were equal to

$$\delta_{11} = + 443.6 \tfrac{1}{E}, \quad \delta_{12} = - 409.9 \tfrac{1}{E}, \quad \delta_{13} = - 566.3 \tfrac{1}{E},$$

$$\delta_{14} = + 187.1 \tfrac{1}{E}, \quad \delta_{15} = + 153.4 \tfrac{1}{E}, \quad \delta_{16} = - 154.8 \tfrac{1}{E}, \quad \delta_{17} = 0;$$

$$\delta_{22} = + 1201.0 \tfrac{1}{E}, \quad \delta_{23} = + 1283.0 \tfrac{1}{E}, \quad \delta_{24} = - 594.8 \tfrac{1}{E},$$

$$\delta_{25} = - 473.8 \tfrac{1}{E}, \quad \delta_{26} = + 110.7 \tfrac{1}{E}, \quad \delta_{27} = - 122.9 \tfrac{1}{E};$$

$$\delta_{33} = + 1907.0 \tfrac{1}{E}, \quad \delta_{34} = - 696.8 \tfrac{1}{E}, \quad \delta_{35} = - 553.2 \tfrac{1}{E},$$

$$\delta_{36} = + 189.9 \tfrac{1}{E}, \quad \delta_{37} = - 210.8 \tfrac{1}{E};$$

$$\delta_{44} = + 660.8 \tfrac{1}{E}, \quad \delta_{45} = + 408.8 \tfrac{1}{E}, \quad \delta_{46} = 0, \quad \delta_{47} = 0;$$

$$\delta_{55} = + 462.0 \tfrac{1}{E}, \quad \delta_{56} = 0, \quad \delta_{57} = 0;$$

$$\delta_{66} = + 121.3 \tfrac{1}{E}, \quad \delta_{67} = 0;$$

$$\delta_{77} = + 452.3 \tfrac{1}{E};$$

$$\Delta_{1p} = + 187,500 \tfrac{1}{E}, \quad \Delta_{2p} = - 266,200 \tfrac{1}{E}, \quad \Delta_{3p} = - 367,100 \tfrac{1}{E},$$

$$\Delta_{4p} = + 116,300 \tfrac{1}{E}, \quad \Delta_{5p} = + 92,360 \tfrac{1}{E}, \quad \Delta_{6p} = - 92,200 \tfrac{1}{E},$$

$$\Delta_{7p} = - 28,970 \tfrac{1}{E}.$$

Solving system of equations (1), we obtain the following values of the unknowns: $X_1 = -95.24$, $X_3 = +94.68$, $X_5 = +9.64$, $X_2 = +68.74$, $X_4 = +6.71$, $X_6 = +427.6$, $X_7 = +126.9$. The numerical values $X_4 = +6.71$ kg and $X_5 = +9.64$ kg show that the forces in the arc (annular) members are insignificant, which is a consequence of the small transverse displacements of the end points of the cantilevers (deflections) of the radial trusses and warrants simplification of the design scheme and main system, which is obtained by reducing the number of arc members considered (see Figs. 3 and 6).

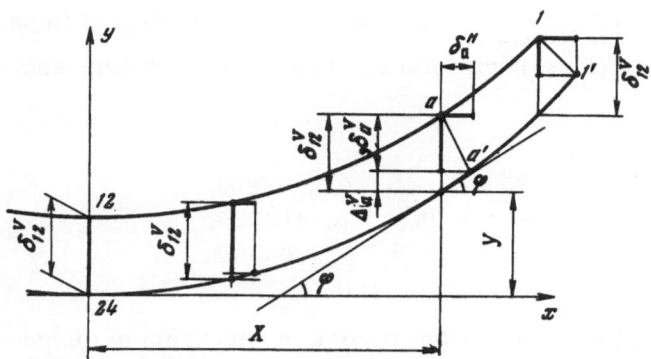

Fig. 8. Diagram of parallel displacement of the mirror surface.

The total (design) values of the forces in the bars of the system are determined by the relation

$$N_\Sigma = N_p + \Sigma \overline{X}_i X_i. \tag{5}$$

Displacements (deflections) were determined at six points of the radial truss (1, 3, 5, 7, 9, and 12) in the main system (see Figs. 3 and 6), i.e., for the isolated part of the mirror having two supports (at points 1 and 5 or, on changing to a plane diametral truss, at points 3 and 16, which are denoted also by letters a and n).

The vertical and horizontal components of the indicated displacements were determined separately. In so doing, the vertical components of the displacement of the main support points (1 and 5) were taken to be equal to zero, since they are, so to speak, a migratory motion for the entire isolated part of the mirror as a whole without distortion of the form.

The vertical displacements were determined by the formulas

$$\delta_i^V = \left(\Sigma \overline{V}_i N_\Sigma \frac{l}{EF} \right)_{\mathrm{rad.t}}, \tag{6}$$

where $\overline{V}_i$ are the forces in the bars of the radial truss resulting from a unit vertical force applied at the point at which the displacement is sought; N_Σ are the design forces in the same bars. To determine the displacements at points 5, 7, and 9 the isolated part of the mirror (see Fig. 6a) was loaded by four unit forces at points 5, 5 and 5', 5', and the sum $\left(\Sigma V_i N_\Sigma \frac{l}{EF} \right)_{\mathrm{rad.t}}$ was taken only for one radial truss. To determine the displacement of the central point (12) a unit force was applied at it and summation $\left(\Sigma \overline{V}_i N_\Sigma \frac{l}{EF} \right)_{4\ \mathrm{rad.t}}$ was carried out over the entire isolated part of the mirror, i.e., with respect to four radial trusses.

The numerical values of the vertical components of the displacements are equal to the following values (in mm):

$$
\begin{aligned}
\delta_{12}^V &= 0.0025 & \delta_5^V &= 0.0185, \\
\delta_9^V &= 0.0094, & \delta_{3\,(a)}^V &= 0.0028, \\
\delta_7^V &= -0.0001, & \delta_1^V &= 0.0066.
\end{aligned}
\tag{7}
$$

The horizontal components of displacement were determined at points 1, 3, 5, and 7 by formulas similar to (6), where $\overline{V}_i$ is replaced by H_i, the forces in the bars from the horizontal unit force. Summation $\left(\Sigma H_t N_\Sigma \frac{l}{EF}\right)_{\text{rad.t}}$ was carried out with respect to one radial truss (Fig. 6b). The numerical values of the horizontal components of displacements proved to be equal to (in mm):

$$
\begin{aligned}
&\delta_{12}^H = 0, &&\delta_5^H = 0.0105, \\
&\delta_9^H = \sim 0, &&\delta_{3(a)}^H = 0.0039, \\
&\delta_7^H = 0.0049, &&\delta_1^H = 0.0053.
\end{aligned}
\tag{8}
$$

As we see from (7), the values of the vertical components of displacement at the points in question proved to be different (although in principle we could attain their equality), since the variation of these values was due to a change of the cross sections of a certain limited number of members, namely: the bars of the pyramid 0-13 and 0-17 (see Fig. 3), the counterbalance beam B-B, the knee braces 0-t (see Fig. 4), and the struts 1-2, 3-4, 5-6, etc. (see Fig. 3).

However, if we take into account the horizontal components of the indicated displacements, the values of the vertical components of the corresponding points cannot be alike, but should satisfy the condition of arrangement of the displaced points on the initial parabolic curve forming the mirror.

The values of the vertical displacements should satisfy the following relations (Fig. 8):

$$
_{\text{eq}}\delta_a^V = \delta_{12}^V - \Delta_a^V,
\tag{9}
$$

where

$$
\Delta_a^V = \frac{\delta_a^H}{2F}\, X.
\tag{10}
$$

In our case, the paraboloidal surface of the mirror is determined by the equation of the generatrix of a parabola:

$$
y = \frac{x^2}{4F},
\tag{11}
$$

where F = 3.25 m and y = x²/13.

From (9), (10), and (11) we have

$$
_{\text{eq}}\delta_a^V = \delta_{12}^V - \frac{\delta_a^H}{6.5}\, X.
\tag{12}
$$

Substituting the values of δ_i^H from (8) into (12) with consideration of the distance of the point in question from the center of the mirror X, we have the following equivalent values of vertical displacements (in mm)

$$
\begin{aligned}
&_{\text{eq}}\delta_1^V = -0.00055, &&_{\text{eq}}\delta_7^V = 0.00130, \\
&_{\text{eq}}\delta_3^V = 0.00068, &&_{\text{eq}}\delta_9^V = 0.00250, \\
&_{\text{eq}}\delta_5^V = -0.00125, &&_{\text{eq}}\delta_{12}^V = 0.00250.
\end{aligned}
\tag{13}
$$

The difference between the corresponding values of displacements according to (7) and (13) is a distortion of the form of the mirror. The maximum value of the indicated difference in our case occurs for point 5 and amounts to $\Delta_s \approx 0.017$ mm.

Deformation of parabolic mirrors manufactured with respect to one force diagram with the same suspension structure and with the same load factor are proportional to the square * of the ratio of their diameters [3]. Consequently, the maximum deflection distorting the form of the mirror at point 1 (see Fig. 8) for a 70-m diameter mirror is equal to

$$(\Delta_5)_{\phi 70} = \Delta_5 \left(\frac{70}{7.5}\right)^2 = 0.017 \cdot 9.32^2 = 1.50 \text{ mm}.$$

However, if we take into account that the deflection at this point for the mirror in the vertical position is about twice the deflection for the horizontal position (symmetrical loading), we can take for the maximum value of deformation for the 70-m diameter mirror (at its edge):

$$(\Delta_1)_{\phi 70}^{max} \approx 3 \text{ mm}.$$

If we take a mean-square value of the distortion of the shape of the mirror equal to

$$\frac{1}{2.6}(\Delta_1)_{\phi 70}^{max} \approx 1.2 \text{ mm},$$

the 70-m mirror on a multiple supporting structure can be effectively used for operation on the 2-cm wavelength [4], since the deviation of the surface of the mirror from the design (as a consequence of elastic deformations) is

$$\Delta \delta = \frac{\lambda}{\frac{1}{2.6}(\Delta_1)_{\phi 70}^{max}} = \frac{20}{1.2} \approx 16.5.$$

In addition to deformations, the accuracy of the mirror surface is also affected by the errors of its manufacture. Experience in the construction [5] of high-precision parabolic mirrors shows that the maximum manufacturing errors are of the order of

$$\Delta_{er. man} = 1 \cdot 10^{-5} D_{mir}.$$

For D_{mir} = 70 m, $\Delta_{er. man} \approx 0.7$ mm. The standard deviation of manufacture is about 0.3 mm. The mean square of the total error is

$$\sum_{\Delta \text{ m.sq.}} = \sqrt{1.2^2 + 0.3^2} \approx 1.25 \text{ mm}.$$

The results of the calculation of the model are not optimal since, first, the lengths of the cantilevers and spans in the model were not optimally selected (it is necessary to reduce slightly the length of the cantilever and the length of the first extreme span); second, the load factor K (see [3]) in the model was too large (about 4.5) and it can be reduced to 3-3.5. These undertakings can increase the mirror rigidity by a factor of about 1.5.

LITERATURE CITED

1. P. D. Kalachev, in: Radio Telescopes (D. V. Skobel'tsyn, ed.), Consultants Bureau, New York (1966), p. 143.
2. P. D. Kalachev, ibid., p. 161.
3. P. D. Kalachev, in: Wideband Cruciform Radio Telescope Research (D. V. Skobel'tsyn, ed.), Consultants Bureau (1969), p. 52.
4. G. Z. Aizenberg, Ultrashort-Wave Antenna [in Russian], Svyaz'izdat (1957).
5. P. D. Kalachev and A. E. Salomonovich, Trudy FIAN, 17:13 (1963).

---

*We have in mind elastic deformations due to its mirror weight.

RADIALLY SYMMETRIC MULTIPLE SUPPORT
OF PARABOLIC ANTENNAS

P. D. Kalachev

In order to construct a large parabolic antenna (which will henceforth be called a parabolic mirror or, simply, a mirror) which is free to rotate in any direction it is necessary for the mirror to possess a high mechanical rigidity. Increases in rigidity can be achieved by supporting the mirror at several points on a rotating-support device [1, 2].

Assuming the main loads on the parabolic mirror are those due to its own weight, we can represent them in the general case as a combination of symmetric and skew-symmetric loads with respect to the horizontal axis. In the particular case, these loads may often be symmetric (horizontal placement of the aperture plane of the mirror) or skew-symmetric (vertical placement of aperture plane of mirror).

If we also allow for loads due to wind (limiting ourselves to a wind speed of 10 m/sec) we see that for a mirror with high mechanical rigidity, i.e., the design weight per square meter of aperture is on the order of 100-200 kg/m^2 (depending on the mirror dimensions), distortion in the distribution of the total loads introduced by wind loads can be ignored in a first approximation and we can limit ourselves to a simple 8-12% increase in weight loads.

The overwhelming majority of present and future parabolic mirrors are designed according to the radial-symmetric scheme [3, 4]. This is quite justified both from the viewpoint of design (one type of element) and from the viewpoint of simplicity and, consequently, the increased computational accuracy [5].

It would seem that the radial-symmetric scheme of mirror construction with a symmetric (skew-symmetric) load would naturally require the use of a radial-symmetric method of suspending the mirror on a rotating-support device. However, in practice, of the many tens of mirrors able to rotate in any direction only one or two have a radial-symmetric support system. Examples of this type are the mirror of the powerful Australian radio telescope near Sydney [3] and the mirror of the FIAN radio telescope [6] which began operating in 1951.

Usually, a parabolic mirror free to rotate in any direction is suspended by using a two-point support beam, i.e., by employing two support pins on the immediate frame of the mirror. This is explained by the fact that the mirror ought to be able to rotate around the horizontal axis ±90° from the zenith. The familiar design difficulties are met with in realizing this requirement. In particular, because of these difficulties, the mirror of the Australian radio telescope [4] can rotate around the horizontal axis only within the limits of 65°.

In the above-mentioned radio telescopes the mirrors have a radial-symmetric suspension; this is also known as a cantilever arrangement. The cantilever suspension, which is the simplest from the design viewpoint, has the important drawback that the main support elements of

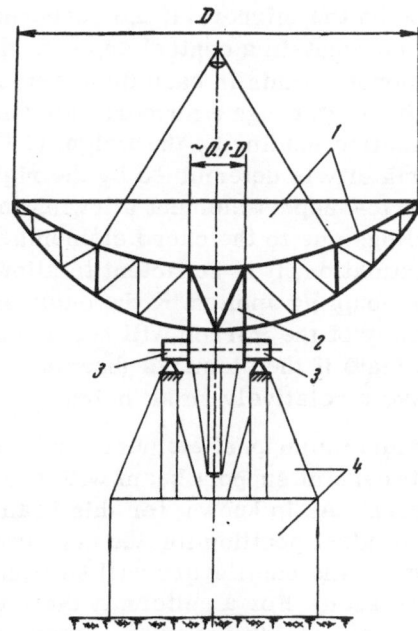

Fig. 1. Simple method of multiple-support suspension of a parabolic mirror (particular case). 1) Mirror; 2) central-support rod; 3) support pins; 4) rotating-support device.

the mirror casing, i.e., the radial elements, are subjected to strong transverse deformations (because they are cantilevered), resulting in a great deal of "sag." However, the cantilever suspension arrangement has one very important advantage. Because it is cyclic, it has the property that for skew-symmetric loading the stresses on the main support elements (in the radial and chordal elements and also in the struts formed by the chords and radial elements) change according to a cyclic law [7], i.e.,

$$N_{r_i} = 2P \sin(2i - 1)\varphi, \qquad (1)$$

where N_{ri} is the load on the i-th radial element acting in the plane of this element and applied to its cantilever end, P is the external vertical load (part of the weight of the mirror) applied to the same end of the radial element, $(2i - 1)\varphi$ is the angle between the plane of the i-th radial element and the plane of the horizontal. Since the stresses in the rods of the radial support element (radial girder) are proportional to the load N_r acting on the element, they will change according to the same law as N_r, i.e., cyclically. The transverse displacements (sags) in the ends of the radial girders are also proportional to the load, and consequently, to the sine of the angle between a given radial element and the plane of the horizontal. On the other hand [7], the distances of the ends of the radial-element cantilevers from the horizontal axis h (length of intersection of mirror aperture plane with plane of the horizontal) are also proportional to the sine of the same angle:

$$h_i = 0.5 D_{\text{mir}} \sin(2i - 1)\varphi. \qquad (2)$$

Thus, the sags in the ends of the radial-element cantilevers are proportional to their distances from the indicated horizontal axis and, consequently, in the deformed state the aperture plane of the mirror is not distorted. This property of the radial-symmetric support system can be successfully used in the design of multiple-support schemes for the suspension of a parabolic mirror. By using a radial-symmetric scheme for the design of the intermediate support we can provide any required number of mirror supports on one surface. In prior schemes, the number of supports on one surface was limited to two. The cyclic properties of the radial-symmetric intermediate design can, of course, only show up under particular loading conditions due to the weight of the mirror.

Figure 1 shows the cantilever method of suspending a parabolic mirror, i.e., to the side of the central support rod of the housing. This method is a particular case of the multiple-support cyclic scheme in which all supports merge to form the central one.

In the general case, single-layer multiple support can be realized using an intermediate radial-symmetric cyclic construction which, in turn, is suspended from a central support rod.

There is a radial support element in the intermediate design for each radial support element on the mirror housing. Therefore, the number of supports is equal to the number of radial

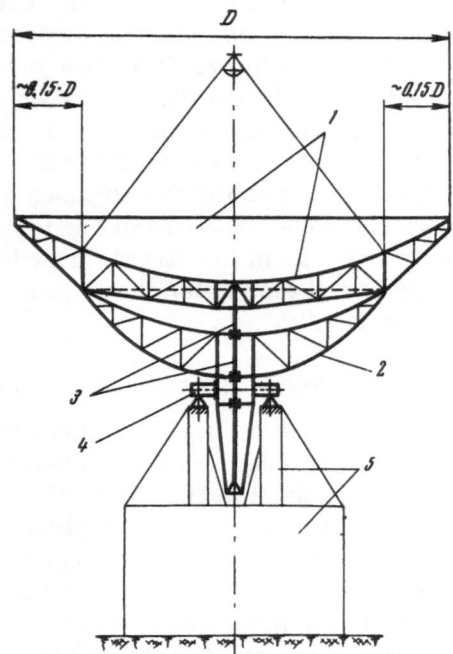

Fig. 2. Multiple support method of suspending a parabolic mirror with central-support rod. 1) Mirror; 2) multiple intermediate-support construction; 3) central support rods; 4) support pins; 5) rotating-support device.

support points on the mirror. If the suspension scheme does not contain a central support, there will be two support points in each diametric section of the mirror (through a support element). For the symmetric loading by the weight of the mirror, the rigidity is determined by the rigidity of one diametric-support element allowing for additional couplings due to the chord elements. To a first approximation, it is sufficient to allow for only one such coupling applied by the outer end; then, the housing of the mirror will become statically indeterminate if the elements (beams) of the outer ring have a relatively small height.

The diametric support element can be treated as a simple two-point support beam with two identical cantilevers. As is known, for this beam there exists an ideal position for the supports at which the sags in the cantilevers will be equal to the sags in the arch. For a uniformly distributed load and a constant rigidity over the length of the beam (diametric element), the distance between supports is roughly equal to $0.56 D_{mir}$. Allowing for the fact that the load at the ends of the cantilever increases, the dimension of the arch section (distance between supports) must increase somewhat and the cantilever must decrease [8] and be about $0.2 D_{mir}$.

If the mirror is intended to operate in the 1- to 2-cm region, then the absolute length of the cantilever must not exceed 5 m. The maximum dimension of the mirror in this case is

$$D_{mir}^{max} \approx \frac{5}{0.2} = 25 \text{ m.}$$

In order to increase the mirror diameter without decreasing the effective area by elastic deformation it is necessary to increase the number of supports in the diametric section of the mirror.

Figure 2 shows a diagram of the multiple-support suspension of a mirror with a radial-symmetric positioning of the supports around a central rod. For such a suspension the mirror contains three supports in each diametric section (with respect to the support elements of the casing) and each diametric element operates as a three-point support beam.

The cross-sectional area of the central rod is chosen so that the vertical displacements of the mirror will be the same at all support points. By taking the absolute dimension of the cantilever as 5 m we obtain as the maximum diameter of a mirror suspended according to the scheme in Fig. 1

$$D_{mir} \approx \frac{5}{0.15} \approx 33 \text{ m.}$$

A further increase in mirror dimensions while maintaining the same rigidity, i.e., without increasing the absolute elastic deformation, results in a complication of the method and construction of the suspension.

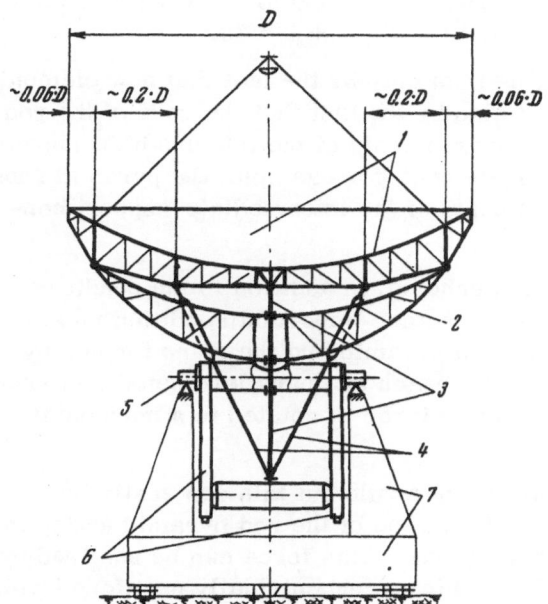

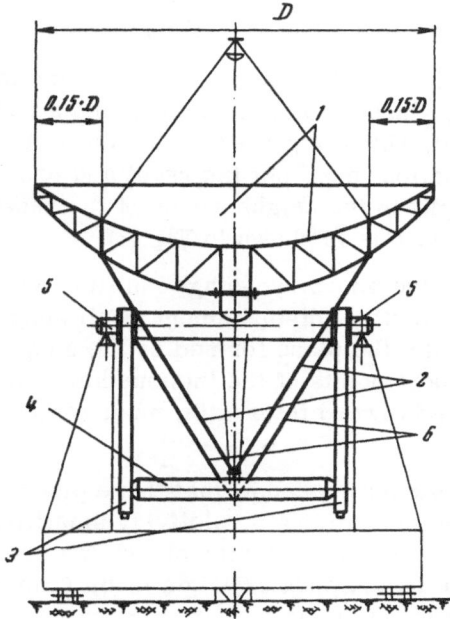

Fig. 3. Multiple-support two-stage scheme of suspending a parabolic mirror with central support rod. 1) Mirror; 2) intermediate-support construction; 3) central-support rod; 4) rod pyramid; 5) support pins; 6) rotation-sector elements with counterweight beam; 7) rotating-support device.

Fig. 4. Multiple-support scheme of suspending a parabolic mirror without intermediate-support construction. 1) Mirror; 2) rod pyramid; 3) rotation sectors; 4) counterweight beam; 5) support pins; 6) suspension of counterweight beam.

Figure 3 shows a multiple-support suspension of a mirror with two-stage radial-symmetric positioning of the supports and with a central support rod. It is not possible to consider this type of mirror suspension as being purely cyclic, since the cyclic suspension for the rod (support) pyramid is provided for by the radial symmetry of the mirror and is insufficient for maintaining the cyclic properties of the intermediate construction of this stress. We shall call this a mixed-cyclic suspension.

The number of support rods for the rod pyramid is equal to the number of radial-support elements in the mirror housing; the number of radial elements in the intermediate-support construction is also equal to the number of radial-support elements of the mirror casing.

The cross-sectional area of the rods in the pyramid, which are all the same, is chosen so that the vertical displacements of all mirror supports are the same. With the same reasoning in mind, the cross-sectional area of the central-support rod is chosen to allow for the elastic deformation of the counterweight beam, which truncates the top of the rod pyramid, and to allow for the elasticity of the rotation-sector elements.

In Fig. 3, five supports are contained in each diametric mirror section (with respect to the support elements). Values for the cantilevers are about $0.06D_{mir}$ for an absolute cantilever dimension of 5 m; the diameter of the reflector (assuming the same rigidity as in the preceding examples), is

$$D_{mir} = \frac{5}{0.06} \approx 80 \text{ m}.$$

If the number of radial-support elements of the mirror is assumed to be 16 (eight diametric elements), the overall number of mirror supports will be 16 · 2 + 1 = 33.

In the given scheme, the construction is complicated not only by the fact that new elements are introduced (rotation sectors and a rod pyramid) but also by the fact that the rods of the rod pyramid, which is placed in the plane of the radial-support elements of the intermediate-support construction, must not intersect and pass into any other element (for example, the pyramid rods must split in the neighborhood of the radial-support elements of the intermediate-support construction; this is shown in Fig. 3).

In the above two-stage multiple-support suspension scheme, in addition to the condition that the vertical displacement of all supports be the same, another important condition must be observed — the plane formed by the support points of the rod pyramid and the plane formed by the support points of the intermediate-support construction, which are parallel when the mirror is pointed toward the zenith, must remain parallel when the mirror is rotated to a horizontal position.

Calculations show that these planes are not parallel* as a rule but this can easily be achieved by applying a relatively small transverse force to the top of the rod pyramid acting in the plane in which the mirror rotates around the horizontal axis. This force can be realized by a torque on the counterweight beam from the balance load which is eccentrically positioned with respect to this girder.

In the schemes considered (Figs. 2, 3), the multiple-support suspension of the mirror is designed so that the main-support elements carrying the weight when the mirror aperture plane is in the vertical position are at a great distance from the central insert at the periphery; as a result, it is necessary to use the intermediate rigid (consequently heavy) support construction.

If we use a rotating-support device with a large horizontal base (similar to the one used in the English radio telescope employing a 76-m parabolic mirror† [9]), multiple-support suspension of the mirror can be realized without using an intermediate support construction. Here, the main attachment of the mirror is between its housing and the central insert.

Figure 4 shows a diagram of the multiple-support suspension of a mirror with radial-symmetric positioning of the supports and with the basic attachment of the mirror at the central insert. This scheme is cyclic. Since the weight of the mirror (for a vertical aperture position) is transmitted by the horizontal tube, the latter must accordingly be fastened in two planes. The number of support rods of the pyramid is equal to the number of radial-support elements of the mirror housing and, consequently, three supports are contained in each diametric mirror section with respect to the support elements. The rigidity of the mirror is on the order of rigidity of a mirror suspended according to the scheme in Fig. 2.

Figure 5 shows another type of multiple-support suspension for a mirror; this type of suspension provides a rigidity which is on the same order as the mirror described in Fig. 3. In this scheme, five supports are contained in each diametric section of the mirror. The scheme is cyclic. The same method as the one used in Fig. 3 is used to ensure that the planes in which lie the reference-arc points formed by the outer and inner rod pyramids are parallel. A drawback to the scheme is that the relatively large horizontal components of the support reactions on the outer rod pyramid cause great deformations in the mirror which cannot be compensated for.

---

*In particular, we can show that when the mirror is rotated around the horizontal axis the considered planes will remain mutually parallel.

† Some drawbacks to these rotating devices are the increased moment of friction when the mirror is rotated along the azimuth and the increased moment of inertia which requires higher drive-motor powers.

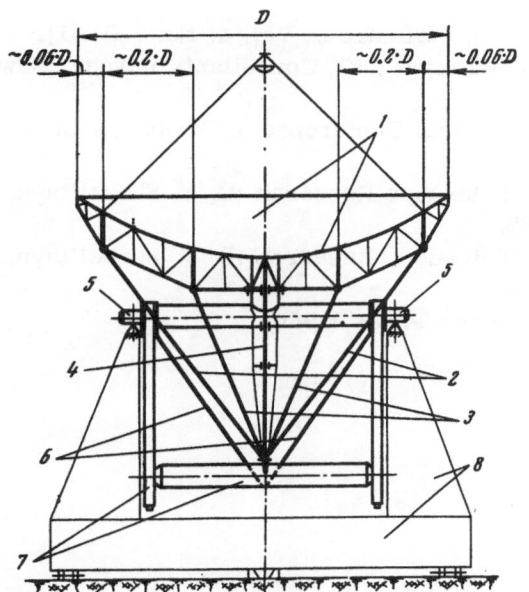

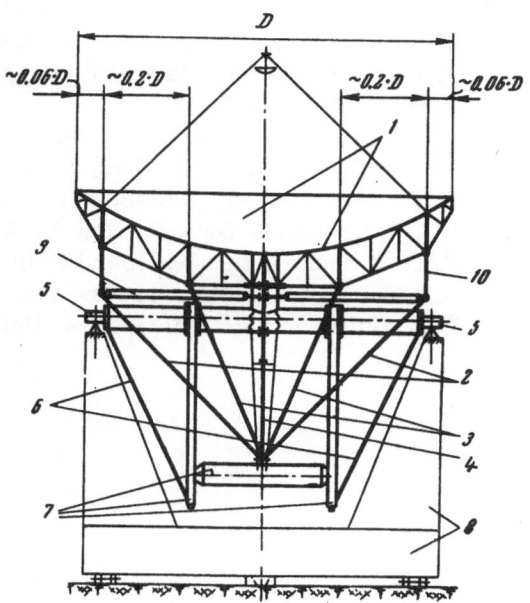

Fig. 5. Two-stage multiple-support method of suspending a parabolic mirror without intermediate support construction. 1) Mirror; 2) external rod pyramid; 3) internal rod pyramids; 4) central-support rod; 5) support pins; 6) suspension of counterweight beam; 7) rotation-sector elements with counterweight beam; 8) rotating-support device.

Fig. 6. Two-stage multiple-support suspension of parabolic mirror with unloaded radial rods.

In order to eliminate this drawback, in the scheme shown in Fig. 6 a weight-distributing system is used which consists of radial rods 9, and the braces 10 are located opposite each rod of the external pyramid. The remaining notation in this figure is the same as in Fig. 5. The number of weight-distributing rods is equal to the number of rods in each of the rod pyramids (and also the number of radial elements in the mirror housing). The sectors which are displaced toward the vertical plane of symmetry make it possible to somewhat decrease the horizontal dimensions of the rotating-support device.

In the last three schemes there is no intermediate support construction; however, a new, quite complex and heavy element appears: the large horizontal beam-tube connecting the rotation sectors and carrying the overall weight of the mirror when the aperture plane is in the vertical position. The horizontal tube must be strongly reinforced at least in two planes: in the plane parallel to the aperture plane of the mirror, and in the vertical plane passing through the support pins.

At this time it is difficult to decide in favor of one of the schemes in Figs. 3-6 since there is no sufficient data (at least for static calculations) for the rigidity of the considered schemes.

In our opinion, it is of fundamental importance that the mirror suspension be multiple-support and radial-symmetric cyclic for the cyclic scheme of mirror housing.

LITERATURE CITED

1. Proc. IRE Australia, No. 9, p. 519 (1959).
2. P. D. Kalachev, in: Radio Telescopes (D. V. Skobel'tsyn, ed.), Consultants Bureau, New York (1966), p. 35.

3.   L. Mohr, Stahlbau, 27(3):62-69 (1958).

4.   P. D. Kalachev and A. E. Salomonovich, Radiotekhn. i Elektron., Vol. 4, No. 3 (1961).

5.   P. D. Kalachev, in: Radio Telescopes (D. V. Skobel'tsyn, ed.), Consultants Bureau, New York (1966), p. 143.

6.   V. V. Vitkevich, Transactions of the Fifth Cosmological Conference on Problems in Radioastronomy, Akad. Nauk SSSR (1956), p. 14.

7.   P. D. Kalachev, in: Wideband Cruciform Radio Telescope Research (D. V. Skobel'tsyn, ed.), Consultants Bureau, New York (1969), p. 52.

8.   P. D. Kalachev, in: Wideband Cruciform Radio Telescope Research (D. V. Skobel'tsyn, ed.), Consultants Bureau, New York (1969), p. 63.

9.   J. G. Bolton, Radio Telescopes, University of Chicago Press (1960).

TOWARD A THEORY OF THE SOLAR WIND

N. M. Dagkesamanskaya and M. V. Konyukov

1. In the investigation of plasma streaming from the sun to a region where the stream mode can be assumed to be laminar ($r > 1.5R_\odot$), the parameter*

$$Q_{e_0}^{-1} = \frac{m_e}{m_i} \frac{m}{m_e + m_i} \frac{4\sqrt{2\pi}n_0 Le^4}{\sqrt{m_e}(kT_0)^{3/2}} \frac{r_0}{v_0} \left(\frac{T_0}{T_{e_0}}\right)^{3/2},$$

which takes into account energy exchange between electrons and ions, has values for which effective equilibrium of electron and ion temperatures is absent and for which it is impossible to consider the flow in an isothermal approximation. The same conclusion follows from direct analysis of the problem of plasma streaming from the sun in the isothermal approximation. This analysis [1] has indicated that the conditions for application of a gas-dynamic approximation are violated at a comparatively small distance from the original level. This means that the collision frequency decreases to values for which temperature equilibrium is impossible and the stream mode ceases to be in equilibrium (study of streaming in the isothermal approximation is particularly necessary as a basis for a model of a stream with a region of frequent collisions of finite radius). Thus, to obtain results satisfactorily describing the properties of a plasma emanating from the sun (the solar wind), the isothermal approximation should be rejected and equations which permit the nonisothermicity of an arbitrary quantity to be described should be used.

Sturrock and Hartley [3] attempted to study a plasma stream in the isothermal approximation, and in order to obtain solutions introduced equivalent polytropic curves. In this work, however, the role of clearly occluded heat sources and the problem of the appropriateness of the hydrodynamic approximation were not studied. The question of the meaning of the approximation when using equivalent polytropic curves was also put aside without being considered. We therefore though it of use to carry out a thorough investigation of a plasma stream in the isothermal approximation and, in particular, to direct special attention to the properties of laminar stream conditions with occluded heat sources.

As an initial system of equations we use the following:

$$vwx^2 = w_0,$$

$$\frac{1}{v}\frac{d}{dx}(v\tau_e) + \left(\frac{er_0}{kT_0}E + a\frac{d\tau_e}{dx}\right) = 0,$$

$$w\frac{dw}{dx} = -\frac{1}{2}\frac{1}{v}\frac{d}{dx}[v(\tau_e + \tau_i)] - \frac{A}{x^2} + \mathrm{Re}_0^{-1}w\frac{d}{dx}\left[x^2\tau_i^{5/2}\left(\frac{dw}{dx} - \frac{w}{x}\right)\right] + \mathrm{Re}_0^{-1}wx\tau_i^{5/2}\left(\frac{dw}{dx} - \frac{w}{x}\right), \qquad (1)$$

---

*The dimensionless variables and parameters used here were introduced in [2].

$$\frac{d}{dx}\left[\frac{w^2}{2}-\frac{A}{x}+\frac{5}{4}(\tau_e+\tau_i)-\mathrm{Pe}_0^{-1}x^2\frac{d\tau_e}{dx}-\mathrm{Re}_0^{-1}x^2\tau_i^{5/2}w\left(\frac{dw}{dx}-\frac{w}{x}\right)\right]=q_ex^2\Phi_e+q_ix^2\Phi_i,$$

$$\frac{d}{dx}\left[\frac{w^2}{2}-\frac{A}{x}+\frac{5}{4}\tau_i-\mathrm{Re}_0^{-1}x^2\tau_i^{5/2}w\left(\frac{dw}{dx}-\frac{w}{x}\right)\right]=\frac{1}{2}\left(\frac{er_0}{kT_0}E+a\frac{d\tau_e}{dx}\right)+Q_{e_0}^{-1}\frac{v^2x^2}{\tau_e^{3/2}}(\tau_e-\tau_i)+q_ix^2\Phi_i.$$
       ⎫
          ⎬ (1)
       ⎭

System (1) is formed from the system of equations of a steady-state, spherically symmetric plasma stream from the sun in the two-fluid approximation [2] as the system of equations for the zero members of an asymptotic series in the small parameter $(m_e/m_i)^{1/2}$

$$\nu_1=\nu(x)+\left(\frac{m_e}{m_i}\right)^{1/2}\nu^{(1)}(x)+\frac{m_e}{m_i}\nu^{(2)}(x)+\cdots,$$

$$w_1=w(x)+\left(\frac{m_e}{m_i}\right)^{1/2}w^{(1)}(x)+\frac{m_e}{m_i}w^{(2)}(x)+\cdots,$$

$$\tau_e=\tau_e(x)+\left(\frac{m_e}{m_i}\right)^{1/2}\tau_e^{(1)}(x)+\frac{m_e}{m_i}\tau_e^{(2)}(x)+\cdots,$$ (2)

$$\tau_i=\tau_i(x)+\left(\frac{m_e}{m_i}\right)^{1/2}\tau_i^{(1)}(x)+\frac{m_e}{m_i}\tau_i^{(2)}(x)+\cdots.$$

The dimensionless parameters of system (1) are expressed by the values of the quantities at the original level and the integrated flow characteristics:

$$\mathrm{Pe}_0^{-1}=\frac{\varkappa_e(T_0)\,r_0}{k/mI_0}\left(\frac{T_{e_0}}{T_0}\right)^{5/2},\qquad \mathrm{Pe}_{0\,(i)}^{-1}=a_i\left(\frac{m_e}{m_i}\right)^{1/2}\mathrm{Pe}_0^{-1}\left(\frac{T_{i_0}}{T_{e_0}}\right)^{5/2},$$

$$\mathrm{Re}_0^{-1}=\frac{4/3\,\mu_i(T_0)\,r_0}{I_0}\left(\frac{T_{i_0}}{T_0}\right)^{5/2},\qquad \mathrm{Re}_{0\,(e)}^{-1}=b_e\left(\frac{m_e}{m_i}\right)^{1/2}\mathrm{Re}_0^{-1}\left(\frac{T_{e_0}}{T_{i_0}}\right)^{5/2},$$ (3)

$$A=\frac{GM_\odot m}{r_0kT_0}\ .$$

At present, data on local powers of heat sources $q_i\Phi_i$ and $q_e\Phi_e$, which are necessary for the analysis of the effect of heat sources on the properties of the forming flow, are absent, and therefore we use in the present work a phenomenological description of the power of the heat sources which conveys only two important characteristics of the sources: the integrated power and the effective dimension. For simplicity, the power of a heat source of this type is considered to depend only on position and acts on the ionic component of the plasma:

$$\Phi_i=\Phi_i(x),\ \ \Phi_e=0.\qquad\qquad(4)$$

In this case the system of equations (1) has a first integral

$$\frac{w^2}{2}+\frac{5}{4}(\tau_e+\tau_i)-\mathrm{Pe}_0^{-1}x^2\tau_e^{5/2}\frac{d\tau_e}{dx}-\mathrm{Re}_0^{-1}x^2w\tau_i^{5/2}\left(\frac{dw}{dx}-\frac{w}{x}\right)=q_i\int_1^x\xi^2\Phi_i(\xi)\,d\xi+\varepsilon_0.\qquad(5)$$

Assuming that the power of the heat source depends only on the coordinate ξ permits us to establish the two characteristics of the heat source without integrating system (1). In fact, to determine q_i from the integrated power we use the equation

$$\varepsilon_\infty-\varepsilon_0=q_i\int_1^\infty\xi^2\Phi(\xi)\,d\xi,\qquad\qquad(6)$$

while the effective dimension is determined from

$$\alpha_0(\varepsilon_\infty-\varepsilon_0)=q_i\int_1^x\xi^2\Phi_i(\xi)\,d\xi,\qquad\qquad(6')$$

where α_0 is that part of the integrated power by which the effective dimension is determined.

In line with contemporary representations of the mechanism of heating of the solar corona, a heat source arises owing to dissipation of mechanical forces, the energy of which is basically carried by ions. Thus, with dissipation the heat should be given off by the ionic components, and its transfer by electrons can be realized either through electron — ion collisions or by induced electrical fields.

In the present work, system (1) with integral (5) is considered in the light of special assumptions about the dimensionless parameters of the problem:

$$\mathrm{Re}_0^{-1} \ll 1, \;\; \mathrm{Pe}_0^{-1} \ll 1, \;\; Q_{e_0}^{-1} \approx 1, \tag{7}$$

$$\mathrm{Re}_0^{-1} \ll 1, \;\; \mathrm{Pe}_0^{-1} \gg 1, \;\; Q_{e_0}^{-1} \approx 1. \tag{7'}$$

2. For parameter values satisfying conditions (7), system of equations (1) has two small parameters: Re_0^{-1} and Pe_0^{-1}, and its solution may be sought in the form

$$
\begin{aligned}
w &= w^{(0)}(x) + \mathrm{Re}_0^{-1} w^{(1)}(x) + \cdots, \\
\tau_e &= \tau_e^{(0)}(x) + \mathrm{Re}_0^{-1} \tau_e^{(1)}(x) + \cdots, \\
\tau_i &= \tau_i^{(0)}(x) + \mathrm{Re}_0^{-1} \tau_i^{(1)}(x) + \cdots.
\end{aligned}
\tag{8}
$$

Considering Re_0^{-1} and Pe_0^{-1} to be of the same order of smallness, we obtain from (1) the following system of equations for the zero members of series (8):

$$
\begin{aligned}
&\nu^{(0)} w^{(0)} x^2 = u_0, \\
&w^{(0)} \frac{dw^{(0)}}{dx} = -\frac{1}{2}\frac{1}{\nu^{(0)}}\frac{d}{dx}\left[\nu^{(0)}(\tau_e^{(0)} + \tau_i^{(0)})\right] - \frac{A}{x^2}, \\
&\frac{d}{dx}\left[\frac{w^{(0)2}}{2} + \frac{5}{4}(\tau_e^{(0)} + \tau_i^{(0)}) - \frac{A}{x}\right] = q_i x^2 \Phi_i(x), \\
&\frac{d\tau_i^{(0)}}{dx} = -\frac{2}{3}\frac{\tau_i^{(0)}}{w^{(0)}}\frac{dw^{(0)}}{dx} - \frac{4}{3}\frac{\tau_i^{(0)}}{x} + \frac{4}{3}\frac{Q_{e_0}^{-1} w_0^2}{x^2 w^{(0)2}\tau_i^{(0)3/2}}(\tau_e^{(0)} - \tau_i^{(0)}) + \frac{4}{3}x^2 q_i \Phi_i,
\end{aligned}
\tag{9}
$$

while the first integral (5) in the same approximation is written in the form

$$\frac{w^{(0)2}}{2} + \frac{5}{4}(\tau_e^{(0)} + \tau_i^{(0)}) - \frac{A}{x} = q_i \int_1^x \xi^2 \Phi_i(\xi)\,d\xi + \varepsilon_0. \tag{10}$$

The system of Eq. (9), after eliminating the density and using integral (10), takes the form

$$
\frac{dw^{(0)}}{dx} = \frac{w^{(0)}}{x} \cdot \frac{-w^{(0)2} + \dfrac{1}{2}\dfrac{A}{x} - q_i x^2 \Phi_i(x) + 2q_i \displaystyle\int_1^x \xi^2 \Phi_i(\xi)\,d\xi + 2\varepsilon_0}{2w^{(0)2} - \dfrac{A}{x} - q_i \displaystyle\int_1^x \xi^2 \Phi_i(\xi)\,d\xi - \varepsilon_0},
$$

$$
\frac{d\tau_i^{(0)}}{dx} = -\frac{2}{3}\frac{\tau_i^{(0)}}{w^{(0)}}\frac{dw^{(0)}}{dx} - \frac{4}{3}\frac{\tau_i^{(0)}}{x} + \frac{4}{3}\frac{Q_{e_0}^{-1} w_0^2}{x^2 \tau_e^{(0)3/2} w^{(0)2}}(\tau_e^{(0)} - \tau_i^{(0)}) + \frac{4}{3} q_i x^2 \Phi_i(x),
$$

$$
\tau_e^{(0)} = \frac{4}{5}\left[\frac{A}{x} + q_i \int_1^x \xi^2 \Phi_i(\xi)\,d\xi - \frac{w^{(0)2}}{2} - \varepsilon_0\right] - \tau_i^{(0)}.
\tag{11}
$$

One of the interesting properties of the motion equation of system (11) is the existence of conditions for which it has no singular saddle-type point. Especially important in this regard is the no-source case. If the numerator and denominator of the motion equation of system (11) vanish, we obtain a system of algebraic equations for determining the singular point:

$$w_*^{(0)^2} - \frac{1}{2}\frac{A}{x_*} = 2\varepsilon_0,$$

$$w_*^{(0)^2} - \frac{1}{2}\frac{A}{x_*} = \frac{1}{2}\varepsilon_0,$$

(12)

where x_* and w_* are the coordinates of the singular point on the plane (x, w), and ε_0 is a constant of integration. It is easily seen that the system of equations (12) is inconsistent for $\varepsilon_0 \neq 0$ and, consequently, in the no-source case, for any departure in temperature, steady-state streaming of plasma from the sun with a smooth transition through the speed of sound is impossible.* If the power of the heat source does differ from zero, then the equation of motion has a singular saddle point. To analyze streaming with a three-dimensional heat source, equations which determine the hydrodynamic velocity and the half-sum of the temperatures are isolated from system (11):

$$\frac{dw^{(0)}}{dx} = \frac{w^{(0)}}{x}\frac{-w^{(0)^2} + \frac{1}{2}\frac{A}{x} - q_i x^3 \Phi_i(x) + 2q_i \int_1^x \xi^2 \Phi_i(\xi)\, d\xi + 2\varepsilon_0}{2w^{(0)^2} - \frac{A}{x} - q_i \int_1^x \xi^2 \Phi_i(\xi)\, d\xi - \varepsilon_0},$$

(13)

$$\tau = \frac{2}{5}\left[q_i \int_1^x \xi^2 \Phi_i(\xi)\, d\xi + \frac{A}{x} - \frac{w^{(0)^2}}{2} + \varepsilon_0\right],$$

where $\tau = (\tau_e^{(0)} + \tau_i^{(0)})/2$. The system of equations (13) agrees with a system of equations describing streaming in the isothermal approximation for analogous conditions, and can easily be interpreted by numerical methods [4]. Analysis of possible streaming modes with a break in temperature can be made utilizing the results of integration of the corresponding isothermal variants. This allows us to make the following conclusions on streaming in an isothermal regime:

a) For heat sources with powers of the form

$$\Phi_i = \frac{1}{x^n}, \quad \Phi_i = e^{-\frac{(x-1)}{2}}, \quad \Phi_i = e^{-\frac{(x-1)^2}{L^2}}$$

(14)

the motion equation of system (13) has a singular saddle point and allows solutions with a smooth transition through the speed of sound at a finite distance from the original level.

b) However, for heat sources of comparatively small effective dimensions, τ quickly increases with distance from the original level, which means an increase in the sum of the electron and ion temperatures. But for a heat source (4) the ion temperature departs from the electron temperature, at least in the region of activity of the heat source, and thus the results of integration will contradict observational data of ion temperature, according to which the temperature declines (we are considering a region behind the temperature maximum of the solar corona). This is one of the reasons preventing the use of the solutions of system (13) to describe the plasma stream even in the isothermal approximation.†

---

*Precisely here does streaming of plasma from the sun most substantially differ from flow through a Laval nozzle. In the latter case even for the no-source case streaming is possible for which the energy of the thermal motion of electrons at $T_{e_0} \gg T_{i_0}$ is converted to hydrodynamic energy of ions.

†System of equations (13) yields solutions for which a gas-dynamic approximation is uniformly

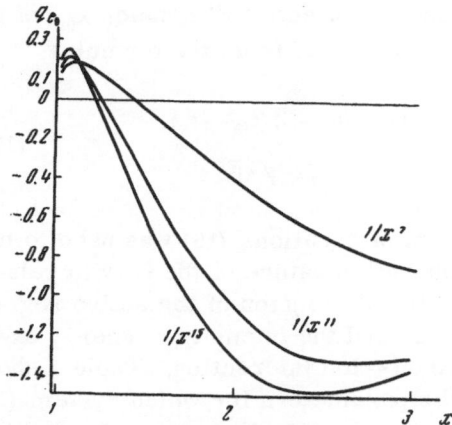

Fig. 1. Dependence of the energy flux q owing to heat conduction on the effective dimension of the heat source (type of source is indicated).

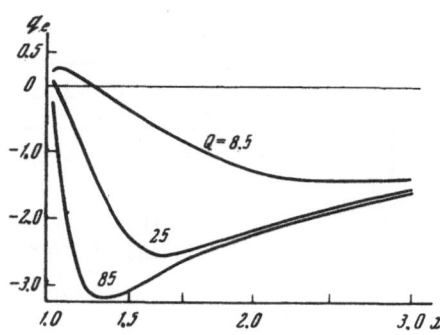

Fig. 2. Dependence of the energy flux q owing to heat conduction on the intensity of electron — ion exchange (parameter of electron exchange is indicated).

3. For parameter values satisfying conditions (7') the solution of system (1) can be presented in the form of a series (8), but now Re_0^{-1} and Pe_0^{-1} cannot be considered as small parameters of like order. The system of equations for the zero approximation with respect to Re_0^{-1} in this case will have the following form:

$$\nu^{(0)}w^{(0)}x^2 = w_0,$$

$$\frac{1}{\nu^{(0)}}\frac{d}{dx}\left(\nu^{(0)}\tau_e^{(0)}\right) + \left(\frac{er_0}{kT_0}E^{(0)} + a\frac{d\tau_e^{(0)}}{dx}\right) = 0,$$

$$w^{(0)}\frac{dw^{(0)}}{dx} = -\frac{1}{2\nu^{(0)}}\frac{d}{dx}\left[(\tau_e^{(0)} + \tau_i^{(0)})\nu^{(0)}\right] - \frac{A}{x^2},$$

$$\frac{d}{dx}\left[\frac{w^{(0)2}}{2} + \frac{5}{4}(\tau_e^{(0)} + \tau_i^{(0)}) - Pe_0^{-1}x^2\tau_e^{(0)5/2}\frac{d\tau_e^{(0)}}{dx} - \frac{A}{x}\right] = q_i x^2 \Phi_i(x),$$

$$\frac{d\tau_i^{(0)}}{dx} = -\frac{2}{3}\frac{\tau_i^{(0)}}{w^{(0)}}\frac{dw^{(0)}}{dx} - \frac{4}{3}\frac{\tau_i^{(0)}}{x} + \frac{4}{3}\frac{Q_{e_n}^{-1}w_0^2}{x^2\tau_e^{(0)3/2}w^{(0)2}}(\tau_e^{(0)} - \tau_i^{(0)}) + \frac{4}{3}q_i\Phi_i x^2. \tag{15}$$

For a heat source with a power depending only on x according to a power law, system (15), after eliminating x and using the first integral of the energy, takes the form

$$\frac{d\tau_e^{(0)}}{dx} = \frac{\dfrac{w^{(0)2}}{2} + \dfrac{5}{4}(\tau_e^{(0)} + \tau_i^{(0)}) - \dfrac{A}{x} + \dfrac{\varepsilon_\infty - \varepsilon_0}{x^{n-3}} - \varepsilon_\infty}{Pe_0^{-1}x^2\tau_e^{(0)5/2}},$$

$$\frac{dw^{(0)}}{dx} = \frac{w^{(0)}}{x}\left[\tau_e^{(0)} + \frac{5}{3}\tau_i^{(0)} - \frac{A}{x} - \frac{2}{3}\frac{Q_{e_3}^{-1}w_0^2}{xw^{(0)2}\tau_e^{(0)3/2}}(\tau_e^{(0)} - \tau_i^{(0)}) - \right.$$

$$\left. - \frac{2(n-3)}{3}\frac{\varepsilon_\infty - \varepsilon_0}{x^{n-3}} - \frac{x}{2}\frac{d\tau_e^{(0)}}{dx}\right]\left[w^{(0)2} - \frac{1}{2}\tau_e^{(0)} - \frac{5}{6}\tau_i^{(0)}\right]^{-1}, \tag{16}$$

$$\frac{d\tau_i^{(0)}}{dx} = -\frac{2}{3}\frac{\tau_i^{(0)}}{w^{(0)}}\frac{dw^{(0)}}{dx} - \frac{4}{3}\frac{\tau_i^{(0)}}{x} + \frac{4}{3}\frac{Q_{e_3}^{-1}w_0^2}{x^2\tau_e^{(0)3/2}w^{(0)2}}(\tau_e^{(0)} - \tau_i^{(0)}) + \frac{4}{3}(n-3)\frac{\varepsilon_\infty - \varepsilon_0}{x^{n-2}},$$

applicable, whereas observations of the ionic distribution function in the region of the earth's orbit indicate that the gas-dynamic approximation breaks down for ions [5].

TABLE 1

A	Pe_0^{-1}	$Q_{e_0}^{-1}$	$\varepsilon_\infty - \varepsilon_0$
4.75	42.53	8.5	4.55
4.75	42.53	25	4.55
4.75	42.53	85	4.55

n	8	11	15
$(x_{0.9} - 1)$	0.59	0.334	0.21

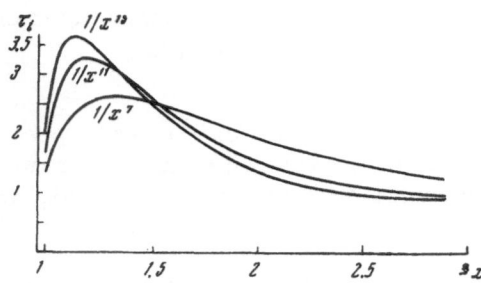

Fig. 3. Dependence of ion temperature on effective dimension of the heat source (source type is indicated).

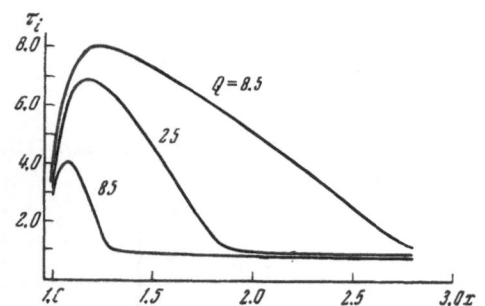

Fig. 4. Dependence of ion temperature on intensity of electron — ion exchange (parameter of electron — ion exchange is indicated).

moreover, q_i and the effective dimension $x_{0.9}$ of the heat source is determined from the formulas

$$q_i = (n-3)(\varepsilon_\infty - \varepsilon_0),$$
$$x_{0.9} = \sqrt[n-3]{10}. \tag{17}$$

The system of equations (16) was used to investigate the change in nature of the flow in relation to the effective dimension of the active region of the heat source and the intensity of energy exchange in electron — ion interaction. Table 1 shows the values of the parameters for which system (16) was integrated, and the effective dimensions of the considered variants are given below. The system of equations (16) was integrated for the parameters indicated in the table for subsonic conditions, and the basic aim was to obtain: a) the behavior of the ion and electron temperatures in the active region of the heat source; b) the nature of formation of the energy flux owing to electron heat conduction; and c) dependence of the solution properties on the effective dimension of the active region of the heat source and the intensity of energy exchange between electrons and ions. The integral curves that will allow us to answer the questions posed are given in Figs. 1-5.

Analyzing these figures, we can draw the following conclusions.

1) In all the considered cases the ion temperature varies nonmonotonically with a maximum in the active region of the heat source. The maximum value of the ion temperature for a given heat source falls off with increasing intensity of electron exchange, and for a given intensity of electron — ion exchange falls off with increasing effective dimension of the heat source.

2) Variation of electron temperature is not great in a region where the variation of ion temperature is substantial. For heat sources with a specified effective dimension an increased intensity of electron — ion exchange leads to a shift from the nonmonotonic variation of the electron temperature with a maximum over the original level to monotonic decrease.

3) Formation of an energy flux owing to electron heat conduction in substance depends on the intensity of electron — ion exchange. Thus, for a comparatively small intensity of electron — ion exchange ($Q_{e_0}^{-1} \approx 8.5$) the hydrodynamic flow increases fairly quickly, which creates a region with a negative energy flow due to heat conduction. At the upper boundary of this region the

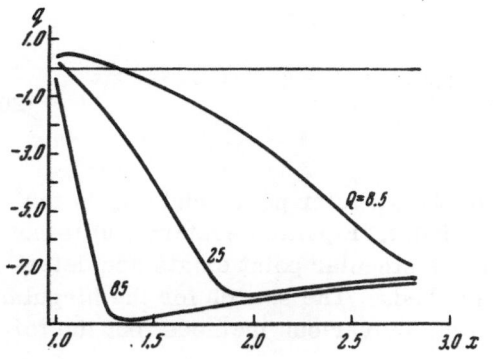

Fig. 5. Dependence of energy flux q owing to heat conductivity on intensity of electron exchange for an energy flux $\varepsilon_\infty = 2.3$.

energy flux due to heat conduction equals zero, and only at this boundary does it become positive and increase, up to a certain level. With increasing intensity of electron exchange the region with negative heat-conducting flow becomes narrower and finally disappears. The electron temperature begins to decrease monotonically at a level x*.

The features of the solutions to the problem of streaming from the sun in the isothermal approximation with heat sources acting in the region of stable laminar flow are found to be in contradiction with the results of observations of behavior of the ion temperature in the corona [6] and of hydrodynamic motions [7]. In fact, to obtain a picture of the variation of the ion temperature, at least qualitatively, communicating the observational data, the original level should be taken between the chromosphere and the maximum-temperature level. However, laminar-flow is most probably not present there, since radar methods indicate a jet structure of the flow even at a level $r \approx 1.5R_\odot$. If the original level is taken at distances exceeding $1.5R_\odot$, the behavior of the ion temperature will contradict observational data. This permits us to make a very important conclusion: in the region where laminar flow exists for plasma streaming from the sun three-dimensional heat sources should be absent.

4) Assuming that in the laminar-flow region heat sources are absent, the first energy integral takes the form

$$\frac{w^{(0)2}}{2} + \frac{5}{4}(\tau_e^{(0)} + \tau_i^{(0)}) - \frac{A}{x} - \mathrm{Pe}_0^{-1}x^2\tau_e^{(0)1/2}\frac{d\tau_e^{(0)}}{dx} = \varepsilon_\infty, \tag{18}$$

where ε_∞ is the asymptotic value of the dimensionless energy flow, while the system of equations describing streaming in these conditions can be written as

$$\frac{d\tau_e^{(0)}}{dx} = \frac{\dfrac{w^{(0)2}}{2} + \dfrac{5}{4}(\tau_e^{(0)} + \tau_i^{(0)}) - \dfrac{A}{x} - \varepsilon_\infty}{\mathrm{Pe}_0^{-1}x^2\tau_e^{(0)1/2}},$$

$$\frac{dw^{(0)}}{dx} = \frac{w^{(0)}}{x}\frac{\dfrac{5}{3}\tau_i^{(0)} + \tau_e^{(0)} - \dfrac{A}{x} - \dfrac{2}{3}\dfrac{Q_{e_n}^{-1}w_0^2}{xw^{(0)2}\tau_e^{(0)1/2}}(\tau_e^{(0)} - \tau_i^{(0)}) - \dfrac{x}{2}\dfrac{d\tau_e^{(0)}}{dx}}{w^{(0)2} - \dfrac{5}{6}\tau_i^{(0)} - \dfrac{1}{2}\tau_e^{(0)}}, \tag{19}$$

$$\frac{d\tau_i^{(0)}}{dx} = -\frac{2}{3}\frac{\tau_i^{(0)}}{w^{(0)}}\frac{dw^{(0)}}{dx} - \frac{4}{3}\frac{\tau_i^{(0)}}{x} + \frac{4}{3}\frac{Q_{e_n}^{-1}w_0^2}{x^2w^{(0)2}\tau_e^{(0)1/2}}(\tau_e^{(0)} - \tau_i^{(0)}).$$

Solutions suitable for interpretation of observed currents should admit a smooth transition through the speed of sound, and since this is possible only at singular saddle points, finding them is one of the basic problems of the theory of streams. It is easily seen that the parameters of the singular point of the motion equation of system (19) satisfy the following system of equations:

$$w_*^{(0)2} - \frac{5}{6}\tau_{i_*}^{(0)} - \frac{1}{2}\tau_{e_*}^{(0)} = 0,$$

$$\frac{5}{3}\tau_{i_*}^{(0)} + \tau_{e_*}^{(0)} - \frac{A}{x} - \frac{2}{3}\frac{Q_{e_n}^{-1}w_{0_*}^2}{x_*w_*^{(0)}\tau_{e_*}^{(0)1/2}}(\tau_{e_*}^{(0)} - \tau_{i_*}^{(0)}) - \frac{x_*}{2}\left(\frac{d\tau_e^{(0)}}{dx}\right)_* = 0, \tag{20}$$

$$\frac{w_*^{(0)^2}}{2} + \frac{5}{4}\,(\tau_{e_*}^{(0)} + \tau_{i_*}^{(0)}) - \frac{A}{x_*} - \mathrm{Pe}_0^{-1}x_*^2\tau_{e_*}^{(0)^{5/2}}\left(\frac{d\tau_{e_*}^{(0)}}{dx}\right)_* = \varepsilon_\infty, \tag{20}$$

where $\omega_*^{(0)}$, $\tau_{e_*}^{(0)}$, $\tau_{i_*}^{(0)}$, x_* and $(d\tau_e^{(0)}/dx)_*$ are parameters of the singular point, and w_{0*} is the initial velocity of the solution passing through the singular point. Equation system (20) is not complete unto itself, since both establishing whether or not a singular point exists and determining its nature is impossible without using numerical methods. The search for the singular point of the motion equation of system (19) can be carried out by various methods, but the following two seem to us to be the simplest.

In the first, for the designated parameters of the problem one constructs a family of integral curves of system (19), corresponding to various w_0, and according to their shape one resolves the question of the presence or absence of a singular point. If the singular point is a saddle point, it can be isolated as a point of intersection of the geometric loci of the maxima of curves $w^{(0)}(x)$ and the maxima of curves $x(w^{(0)})$, which are the solutions of the following system of differential equations:

$$\frac{dx}{dw^{(0)}} = \frac{x}{w^{(0)}}\,\frac{w^{(0)^2} - \frac{5}{6}\tau_i^{(0)} - \frac{1}{2}\tau_e^{(0)}}{\frac{5}{3}\tau_i^{(0)} + \tau_e^{(0)} - \frac{A}{x} - \frac{2}{3}\,\frac{Q_e^{-1}w_0^2}{x^2\tau_e^{(0)^{3/2}}w^{(0)^2}}\,(\tau_e^0 - \tau_i^0) - \frac{x}{2}\frac{d\tau_e^{(0)}}{dx}},$$

$$\frac{d\tau_e^{(0)}}{dw^{(0)}} = \frac{d\tau_e^{(0)}}{dx}\,\frac{dx}{dw^{(0)}}, \tag{21}$$

$$\frac{d\tau_i^{(0)}}{dw^{(0)}} = \frac{d\tau_i^{(0)}}{dx}\,\frac{dx}{dw^{(0)}},$$

where $d\tau_e^{(0)}/dx$ and $d\tau_i^{(0)}/dx$ are taken from the system of equations (19). If the existence is established of a singular saddle point, one seeks by numerical methods a unique solution, subsonic at the original level and supersonic behind the singular point. It is precisely this solution that corresponds to a smooth transition through the speed of sound and is suitable for interpretation of the observed properties of the solar wind.

In the second, one seeks on the (w, x) plane the geometric locus of points at which the numerator and denominator of the motion equation of system (19) vanish, and then using analytical and numerical methods one finds a solution in the interval x from unity to any of the points of this geometric locus. As a result of interpretation we obtain values of $\tau_e^{(0)}$ and $\tau_i^{(0)}$ for x = 1, and this point is taken as the singular point for which the ion and electron temperatures take on the correct values at the original level. *

---

*The geometric locus of points at which the numerator and denominator of the motion equation of system (19) vanish are obtained from (20), specifying $\tau_e^{(0)}$ and eliminating $\tau_i^{(0)}$ and $(d\tau_e^{(0)}/dx)_*$. The singular point is sought in the following way: setting $x_* > 1$ and $\tau_{e_*}^{(0)}$, we find w_*; at the point (x_*, w_*) one finds the excluded values for the derivatives $(dw/dx)_*$ and $(d\tau_i^0/dx)_*$ and with their help the solution, going to x = 1. If the solution found provides the necessary values of $\tau_e^{(0)}(1)$ and $\tau_i^{(0)}(1)$, the original point is considered a singular point. If these values are not provided, the next point from the geometric locus of points at which the numerator and denominator of the motion equation of system (19) vanish is examined.

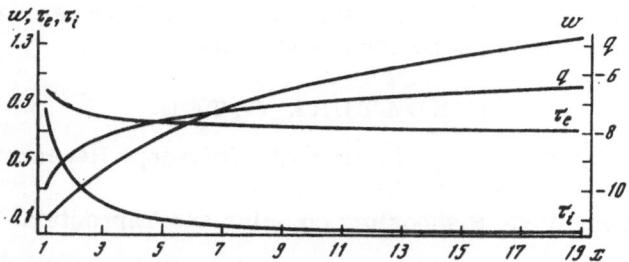

Fig. 6. Singular solution of the no-source problem for $\varepsilon_\infty = 2.3$.

Both in distinguishing the singular point and in obtaining the critical solution to equation system (19) we applied the first method. Concrete calculations were made for parameters

$$\text{Pe}_0^{-1} = 42.53, \quad A = 4.75, \quad Q_{e_0}^{-1} = 8.50, \quad \varepsilon_\infty = 8.00;$$

the results of these calculations are shown in Fig. 6.

In obtaining the critical solution of system (19) criteria were drawn of the usefulness of the hydrodynamic approximation, and in addition the results of these calculations point to the development of a breakdown in the conditions of applicability of the hydrodynamic approximation at comparatively small distances from the original level. *

5. We have investigated the plasma stream in a two-temperature approximation with a significant role played by energy transfer by electron-heat conduction both in the source and no-source regimes. Despite the comparatively large value of the parameter $Q_{e_0}^{-1}$, the solutions indicate a substantial divergence between electron and ion temperatures. This specifically validates investigation of the streaming conditions in the isothermal approximation. Analysis of the concrete results of interpreting plasma streaming from the sun permits us to make a series of conclusions relevant to the properties of the solutions encountered.

1) We cannot consider the heat source which creates the energy flow necessary for the formation of the plasma stream with the properties observed in the region of the earth's orbit to be located within the region of formation of laminar flow.† It follows from this that in investigating a stream using Eq. (1) one must limit the no-source conditions: at the original level there exists an already formed energy flux due to heat conduction that is capable of ensuring the creation of the necessary stream properties.

2) In the no-source case the existence of solutions which smoothly pass through the speed of sound at a singular saddle point is possible. The solutions obtained for a series of values of the problem parameters are such that conditions which preclude use of the gas-dynamic approximation arise at a finite distance from the original level. Thus it is impossible to use the solutions to interpret the properties of the solar wind.

3) If the conditions for using a gas-dynamic approximation within the bounds of the problem considered are violated, this does not mean that collision-free conditions appear at a certain level. This means rather that it is necessary to consider viscosity in problems of plasma

---

*It must be kept in mind that it does not follow from this that there is a collision-free streaming mode. The inclusion of viscosity can significantly change the properties of the solutions, and a final solution demands consideration of the role of viscosity.

† We have radar observations of the solar corona at a level around $1.5R_\odot$, which indicate the presence of a flow of material with mean velocity less than the local velocities, directed away from the sun's surface and toward its surface [7]. From this we specifically can conclude that there is no laminar current already formed at distances of at least up to $1.5R_\odot$.

streaming from the sun, since only after investigating viscous flow can a conclusion as to the appearance of collision-free conditions be considered reliable.

LITERATURE CITED

1. M. V. Konyukov, in: Wideband Cruciform Radio Telescope Research (D. V. Skobel'tsyn, ed.), Consultants Bureau, New York (1969), p. 115.
2. M. V. Konyukov, International Symposium on Solar and Terrestrial Physics [in Russian], Belgrade (1966).
3. P. A. Sturrock and R. E. Hartley, Phys. Rev. Letters, 16:628 (1966).
4. M. V. Konyukov, Geomagnetizm i Aeronomiya, Vol. 7, No. 4 (1967).
5. J. H. Wolfe, R. W. Silva, D. D. McKibbin, and R. H. Mason, The International Union Symposium on Solar and Terrestrial Physics, Belgrade (1966).
6. D. E. Billings and C. G. Lilliegust, Astrophys. J., 137:16 (1963).
7. J. H. Chisholm and J. C. James, Astrophys. J., 140:377 (1964).

A POSSIBLE HYDRODYNAMIC MODEL FOR A FINITE-RADIUS REGION OF FREQUENT COLLISIONS

M. V. Konyukov

1. Observations, from rockets and earth-orbiting satellites, of a continuous flow of plasma from the solar corona [1,2] stimulated interest in Parker's hydrodynamic model of the solar corona and the interplanetary plasma [3]; satisfactory agreement between the magnitudes of the velocities predicted and the observed values has given support to the model. However, an analysis of the conditions in the solar corona necessary to explain an outflow with the observed properties shows, for example, that even under favorable circumstances the heat sources must extend to distances on the order of 5-10R$_\odot$ [4] and this is hard to reconcile with present-day theories on the heating mechanism of the corona.

This difficulty practically disappears if it is assumed that molecular thermal conduction is the basic mechanism of energy transport in at least the region of the flow situated at a small distance above the initial level [5]. The results obtained within the framework of such a model can only be used to describe the outflow provided the gas-dynamic approximation is valid in the whole of the region considered. * Since the gas-dynamic approximation undoubtedly holds near the initial level, the problem of whether gas dynamics can be applied to the whole region occupied by the solar-corona plasma can be investigated as follows. First, one seeks the solution of the kinetic equation in the Enskog — Chapman approximation (to second order in the expansion of the Knudsen number) and then investigates the conditions under which this approximation holds at all points of the flow considered. Results of these calculations permit one to establish which conditions the escaping gas with given properties must satisfy at the initial level if the gas-dynamic approximation is to be uniformly valid over the whole of the considered interval $(1, \infty)$. If it is found that there exist conditions under which the gas-dynamic approximation does not hold, the results obtained in an investigation of the flow in the framework of the Enskog — Chapman approximation will be devoid of meaning and the outflow can only be described by means of a dynamic model with a finite-radius region of frequent collisions.

In order to study this question we restrict ourselves to the steady-state, spherically symmetric outflow of gas in the gravitational field of a star. The system of differential equations describing the outflow in the hydrodynamic approximation may, after the introduction of the

---

*One can easily see that the same problem arises in an investigation of the outflow of an ideal plasma; however, as was shown in [4], a transition to free-molecular conditions is impossible in this case. Below, the same conclusion will be obtained by a different method.

dimensionless variables

$$w = \frac{v}{\sqrt{\frac{kT_0}{m}}}, \quad \tau = \frac{T}{T_0}, \quad x = \frac{r}{r_0} \qquad (1)$$

and the dimensionless parameters

$$\mathrm{Re}_0^{-1} = \frac{4}{4}\frac{\mu_0 r_0}{I_0}, \quad A = \frac{GM_\odot m}{r_0 kT_0}, \quad \varepsilon_\infty = \frac{E_0}{\frac{kT_0}{m}I_0} \qquad (2)$$

be written in the form

$$\mathrm{Re}_0^{-1}x^2\tau^n w^2\frac{d^2w}{dx^2} + \mathrm{Re}_0^{-1}x^2\tau^{n-1}w^2\left(\frac{2\tau}{x} + n\frac{d\tau}{dx}\right)\left(\frac{dw}{dx} - \frac{w}{x}\right) - (w^2 - \tau)\frac{dw}{dx} - w\frac{d\tau}{dx} - w\frac{A}{x^2} + 2\frac{w\tau}{x} = 0,$$

$$\mathrm{Pr}^{-1}\mathrm{Re}_0^{-1}x^2\tau^n\frac{d\tau}{dx} + \mathrm{Re}_0^{-1}x^2\tau^n w\left(\frac{dw}{dx} - \frac{w}{x}\right) - \frac{w^2}{2} - \frac{5}{2}\tau + \frac{A}{x} + \varepsilon_\infty = 0, \qquad (3)$$

where Pr^{-1} is the reciprocal of the Prandtl number and n is the exponent which determines the dependence of the coefficient of viscosity and the thermal conductivity on the temperature:

$$\mu = \mu_0\left(\frac{T}{T_0}\right)^n, \quad \varkappa = \varkappa_0\left(\frac{T}{T_0}\right)^n. \qquad (4)$$

Choosing the following definition for the local Knudsen number

$$\alpha = \frac{\lambda}{r} \qquad (5)$$

and taking into account the fact that

$$\lambda = c\frac{\mu_0\tau^n}{v\tau^{1/2}}, \qquad (6)$$

we obtain the dependence of α on the hydrodynamic parameters and the distance

$$\alpha = \alpha_0 w\tau^{n-\frac{1}{2}} x, \qquad (7)$$

where, in the transition from (5) to (7), we have used the first integral of the system of equations of gas dynamics

$$vwx^2 = w_0. \qquad (8)$$

We note that α may be expressed in the form

$$\alpha = \alpha_{01}\,M\,\mathrm{Re}^{-1}, \qquad (7')$$

where $M = w/\sqrt{\gamma\tau}$ and $\mathrm{Re}^{-1} = \mathrm{Re}_0^{-1}\tau^n x$ is the local Mach number and the reciprocal of the Reynolds number for the flow under consideration [6].

One can easily see that in order to establish when the gas-dynamic approximation is valid, one must solve the system of equations (3), or at least determine the asymptotic behavior of its solutions at large values of x. However, even the determination of the asymptotic behavior of the solutions at great distances for arbitrary values of the parameter Re_0^{-1} is a problem that is difficult to solve; therefore, in investigating this problem by analytic means, we shall restrict ourselves to a consideration of conditions at the initial level under which the values of the reciprocal of the Reynolds number satisfy the inequality

$$\mathrm{Re}_0^{-1} \ll 1. \tag{9}$$

2. If the conditions at the initial level are such that (9) is satisfied, this inequality being so strong that $\mathrm{Pe}_0^{-1} = \mathrm{Pr}^{-1}$, Re_0^{-1} is much less than unity, the solution of the system of equations (3) may be found as an asymptotic series in Re_0^{-1}

$$
\begin{aligned}
w &= w^{(0)}(x) + \mathrm{Re}_0^{-1} w^{(1)}(x) + \mathrm{Re}_0^{-2} w^{(2)}(x) + \cdots, \\
\tau &= \tau^{(0)}(x) + \mathrm{Re}_0^{-1} \tau^{(1)}(x) + \mathrm{Re}_0^{-2} \tau^{(2)}(x) + \cdots,
\end{aligned}
\tag{10}
$$

where $w^{(0)}, w^{(1)}, \ldots, \tau^{(0)}, \tau^{(1)}, \ldots$ are the functions of x to be determined.

Substituting (10) into (3) and setting the coefficients of the same powers of Re_0^{-1} equal to zero, we obtain equations for the determination of $w^{(0)}, w^{(1)}, \ldots, \tau^{(0)}, \tau^{(1)}, \ldots$:

$$
\left.
\begin{aligned}
&(w^{(0)2} - \tau^{(0)}) \frac{dw^{(0)}}{dx} + w^{(0)} \frac{d\tau^{(0)}}{dx} + w^{(0)} \frac{A}{x^2} - 2 \frac{w^{(0)} \tau^{(0)}}{x} = 0, \\
&\frac{w^{(0)2}}{2} + \frac{5}{2} \tau^{(0)} - \frac{A}{x} = \varepsilon_\infty;
\end{aligned}
\right\}
\tag{11'}
$$

$$
\left.
\begin{aligned}
&(w^{(0)2} - \tau^{(0)}) \frac{dw^{(1)}}{dx} + w^{(0)} \frac{d\tau^{(1)}}{dx} + \left(2 w^{(0)} \frac{dw^{(0)}}{dx} + \frac{d\tau^{(0)}}{dx} + \frac{A}{x^2} - \frac{2\tau^{(0)}}{x}\right) w^{(1)} - \\
&- \left(\frac{dw^{(0)}}{dx} + \frac{2 w^{(0)}}{x}\right) \tau^{(1)} = x^2 w^{(0)2} \tau^{(0)n} \frac{d^2 w^{(0)}}{dx^2} + x^2 w^{(0)2} \tau^{(0)(n-1)} \left(\frac{2\tau^{(0)}}{x} + n \frac{d\tau^{(0)}}{dx}\right)\left(\frac{dw^{(0)}}{dx} - \frac{w^{(0)}}{x}\right), \\
&w^{(0)} w^{(1)} + \frac{5}{2} \tau^{(1)} = \mathrm{Pr}^{-1} x^2 \tau^{(0)n} \frac{d\tau^{(0)}}{dx} + x^2 \tau^{(0)n} w^{(0)}\left(\frac{dw^{(0)}}{dx} - \frac{w^{(0)}}{x}\right);
\end{aligned}
\right\}
\tag{11''}
$$

$$
\left.
\begin{aligned}
&(w^{(0)2} - \tau^{(0)}) \frac{dw^{(2)}}{dx} + w^{(0)} \frac{d\tau^{(2)}}{dx} + \left(2 w^{(0)} \frac{dw^{(0)}}{dx} + \frac{d\tau^{(0)}}{dx} + \frac{A}{x^2} - \frac{2\tau^{(0)}}{x}\right) w^{(2)} - \\
&- \left(\frac{dw^{(0)}}{dx} + \frac{2 w^{(0)}}{x}\right) \tau^{(2)} = x^2 w^{(0)2} \tau^{(0)n} \left[\frac{2}{w^{(0)}} \frac{d^2 w^{(0)}}{dx^2} w^{(1)} + \frac{n}{\tau^{(0)}} \frac{d^2 w^{(0)}}{dx^2} \tau^{(1)} + \frac{d^2 w^{(1)}}{dx^2}\right] + \\
&+ x^2 w^{(0)2} \tau^{(0)(n-1)} \Bigg\{ 2 \left(\frac{2\tau^{(0)}}{x} + n \frac{d\tau^{(0)}}{dx}\right)\left(\frac{dw^{(0)}}{dx} - \frac{w^{(0)}}{x}\right)\frac{w^{(1)}}{w^{(0)}} + (n-1)\left(\frac{2\tau^{(0)}}{x} + \right. \\
&+ n \frac{d\tau^{(0)}}{dx}\right)\left(\frac{dw^{(0)}}{dx} - \frac{w^{(0)}}{x}\right)\frac{\tau^{(1)}}{\tau^{(0)}} + \left(\frac{2\tau^{(0)}}{x} + n \frac{d\tau^{(0)}}{dx}\right)\left(\frac{dw^{(1)}}{dx} - \frac{w^{(1)}}{x}\right) + \left(\frac{dw^{(0)}}{dx} - \frac{w^{(0)}}{x}\right)\left(\frac{2\tau^{(1)}}{x} + n \frac{d\tau^{(1)}}{dx}\right)\Bigg\} - \\
&- (2 w^{(0)} w^{(1)} - \tau^{(1)}) \frac{dw^{(1)}}{dx} - w^{(1)2} \frac{dw^{(0)}}{dx} - w^{(1)} \frac{d\tau^{(1)}}{dx} + 2 \frac{w^{(1)} \tau^{(1)}}{x}, \\
&w^{(0)} w^{(2)} + \frac{5}{2} \tau^{(2)} = \mathrm{Pr}^{-1} x^2 \tau^{(0)n} \left(\frac{d\tau^{(1)}}{dx} + n \frac{d\tau^{(0)}}{dx}\frac{\tau^{(1)}}{\tau^{(0)}}\right) + \\
&+ x^2 \tau^{(0)n} w^{(0)} \left[n\left(\frac{dw^{(0)}}{dx} - \frac{w^{(0)}}{x}\right)\frac{\tau^{(1)}}{\tau^{(0)}} + \left(\frac{dw^{(0)}}{dx} - \frac{w^{(0)}}{x}\right)\frac{w^{(1)}}{w^{(0)}} + \left(\frac{dw^{(1)}}{dx} - \frac{w^{(1)}}{x}\right)\right] - \frac{w^{(1)2}}{2}, \text{ etc.}
\end{aligned}
\right\}
\tag{11'''}
$$

The system of equations (11'), which determines the zeroth approximation in Re_0^{-1}, is identical to the equations that describe a steady–state, spherically symmetric flow of an ideal gas in the gravitational field of a star and it may be integrated in a closed form [7]. The system of first integrals of the equations of (11') found by integration can be written in the form

$$
\begin{aligned}
\frac{1}{x^2} &= \frac{w^{(0)}}{w_0}\left[\frac{2}{5}\left(\varepsilon_\infty - \frac{w^{(0)2}}{2} + \frac{A}{x}\right)\right]^{3/2}, \\
\tau^{(0)} &= \frac{2}{5}\left(\varepsilon_\infty - \frac{w^{(0)2}}{2} + \frac{A}{x}\right), \\
v^{(0)} &= \frac{w_0}{w^{(0)} x^2},
\end{aligned}
\tag{12}
$$

where w_0 and ε_∞ are integration constants. Explicit expressions for $w^{(0)}(x)$ and $\tau^{(0)}(x)$, which are required for the integration of the systems of linear differential equations (11"), (11'''), etc., can only be obtained in a neighborhood of a point.* In particular, by iteration one can easily find an asymptotic representation for $w^{(0)}(x)$ and $\tau^{(0)}(x)$ which is valid for large values of x:

$$w^{(0)2} = 2\left[\varepsilon_\infty + \frac{A}{x} - \frac{5}{2}\left(\frac{w_0}{\sqrt{2\varepsilon_\infty}}\right)^{2/5}\frac{1}{x^{4/5}} + 0\left(\frac{1}{x^{7/5}}\right)\right],$$

$$w^{(0)} = \sqrt{2\varepsilon_\infty}\left[1 + \frac{1}{2}\frac{A}{\varepsilon_\infty}\frac{1}{x} - \frac{5}{4}\left(\frac{w_0}{\sqrt{2\varepsilon_\infty}}\right)^{2/5}\frac{1}{\varepsilon_\infty}\frac{1}{x^{4/5}} - \frac{1}{8}\frac{A^2}{\varepsilon_\infty^2}\frac{1}{x^2} + 0\left(\frac{1}{x^{7/5}}\right)\right], \tag{13}$$

$$\tau^{(0)} = \left(\frac{w_0}{\sqrt{2\varepsilon_\infty}}\right)^{2/5}\frac{1}{x^{4/5}} + 0\left(\frac{1}{x^{7/5}}\right)$$

(hypersonic asymptotic behavior);

$$w^{(0)} = \frac{w_0}{\left(\frac{2}{5}\varepsilon_\infty\right)^{3/2}}\frac{1}{x^2} - \frac{3}{2}\frac{w_0}{\left(\frac{2}{5}\varepsilon_\infty\right)^{3/2}}\frac{A}{\varepsilon_\infty}\frac{1}{x^3} + 0\left(\frac{1}{x^4}\right),$$

$$\tau^{(0)} = \frac{2}{5}\varepsilon_\infty + \frac{2}{5}\frac{A}{x} + 0\left(\frac{1}{x^4}\right) \tag{14}$$

(subsonic asymptotic behavior).†

Before turning to the determination of the first approximation, we carry out a number of transformations in the system of equations (11"), after which it can be written in the form

$$\left(\frac{3}{5}w^{(0)2} - \tau^{(0)}\right)\frac{dw^{(1)}}{dx} + \left[2w^{(0)}\frac{dw^{(0)}}{dx} + \frac{d\tau^{(0)}}{dx} - \frac{2\tau^{(0)}}{x} + \frac{A}{x^2} + \frac{4}{5}\frac{w^{(0)2}}{x}\right]w^{(1)} =$$

$$= x^2w^{(0)2}\tau^{(0)n}\frac{d^2w^{(0)}}{dx^2} + x^2\tau^{(0)(n-1)}w^{(0)2}\left(\frac{2\tau^{(0)}}{x} + n\frac{d\tau^{(0)}}{dx}\right)\left(\frac{dw^{(0)}}{dx} - \frac{w^{(0)}}{x}\right) -$$

$$- \frac{2}{5}w^{(0)}\frac{d}{dx}\left[\text{Pr}^{-1}x^2\tau^{(0)n}\frac{d\tau^{(0)}}{dx} + x^2\tau^{(0)n}\left(w^{(0)}\frac{dw^{(0)}}{dx} - \frac{w^{(0)2}}{x}\right)\right] + \tag{15}$$

$$+ \frac{2}{5}\left(\frac{dw^{(0)}}{dx} + \frac{2w^{(0)}}{x}\right)\left[\text{Pr}^{-1}x^2\tau^{(0)n}\frac{d\tau^{(0)}}{dx} + x^2\tau^{(0)n}\left(w^{(0)}\frac{dw^{(0)}}{dx} - \frac{w^{(0)2}}{x}\right)\right],$$

$$\tau^{(1)} = -\frac{2}{5}w^{(0)}w^{(1)} + \frac{2}{5}\text{Pr}^{-1}x^2\tau^{(0)n}\frac{d\tau^{(0)}}{dx} + \frac{2}{5}x^2\tau^{(0)n}w^{(0)}\left(\frac{dw^{(0)}}{dx} - \frac{w^{(0)}}{x}\right).$$

Using the expressions (13) and (14) for $w^{(0)}$ and $\tau^{(0)}$, one can easily calculate the coefficients that occur in Eqs. (15). By means of them Eqs. (15) can be reduced to the following form:

$$\frac{dw^{(1)}}{dx} + \frac{4}{3}\frac{1}{x}\left[1 + 0\left(\frac{1}{x}\right)\right]w^{(1)} = \frac{2(n-3)}{3}\left(\frac{w_0}{\sqrt{2\varepsilon_\infty}}\right)^{\frac{2n}{3}}(2\varepsilon_\infty)^{1/2}\frac{1}{x^{4n/3}}\left[1 + 0\left(\frac{1}{x}\right)\right],$$

$$\tau^{(1)} = -\frac{2}{5}\sqrt{2\varepsilon_\infty}\left[1 + 0\left(\frac{1}{x}\right)\right]w^{(1)} - \frac{4\varepsilon_\infty}{5}\left(\frac{w_0}{\sqrt{2\varepsilon_\infty}}\right)^{\frac{2n}{3}}\frac{1}{x^{\frac{4n}{3}-1}}\left[1 + 0\left(\frac{1}{x}\right)\right] \tag{16}$$

(supersonic flow);

---

*Serious difficulties in the construction of the solutions may occur in the neighborhood of critical points of the system (11'). This must be borne in mind, in particular, in the construction of solutions in the neighborhood of a saddle-type critical point, which is present in the case of an outflow with heat sources. In such a case Lighthill's method should be used [8].

†Neither the form of Eqs. (11') nor their solutions (12)-(14) depend on the parameter n; this cannot be said of the local Reynolds and Knudsen numbers.

$$\frac{dw^{(1)}}{dx} + \frac{2}{x}\left[1 + 0\left(\frac{1}{x}\right)\right]w^{(1)} = \frac{3+2n}{5}\,\mathrm{Pr}^{-1}\frac{w_0}{\left(\frac{2}{5}\,\varepsilon_\infty\right)^{\frac{3}{2}-n}}\cdot\frac{A^2}{\varepsilon_\infty}\,\frac{1}{x^4}\left[1 + 0\left(\frac{1}{x}\right)\right],$$

$$\tag{17}$$

$$\tau^{(1)} = -\frac{4}{25}\,\mathrm{Pr}^{-1}\left(\frac{2}{5}\,\varepsilon_\infty\right)^n A\left[1 + 0\left(\frac{1}{x}\right)\right]w^{(1)} - \frac{1}{5}\frac{w_0}{\left(\frac{2}{5}\,\varepsilon_\infty\right)^{3/2}}\frac{1}{x^2}\left[1 + 0\left(\frac{1}{x}\right)\right]$$

(subsonic flow).

By solving the system of equations (16) and (17), one can find the first approximation $w^{(1)}$ and $\tau^{(1)}$ for supersonic and subsonic flows:

$$w^{(1)} = \frac{2(n-3)}{3-4n}\left(\frac{w_0}{\sqrt{2\varepsilon_\infty}}\right)^{\frac{2n}{3}}\frac{\sqrt{2\varepsilon_\infty}}{x^{\frac{4n}{3}-1}}\left[1 + 0\left(\frac{1}{x}\right)\right],$$

$$\tag{18}$$

$$\tau^{(1)} = -\frac{4}{5}\frac{4-3n}{7-4n}\,\varepsilon_\infty\left(\frac{w_0}{\sqrt{2\varepsilon_\infty}}\right)^{\frac{2n}{3}}\frac{1}{x^{\frac{4n}{3}-1}}\left[1 + 0\left(\frac{1}{x}\right)\right]$$

(supersonic flow);

$$w^{(1)} = -\frac{3+2n}{5}\,\mathrm{Pr}^{-1}\frac{w_0}{\left(\frac{2}{5}\,\varepsilon_\infty\right)^{\frac{3}{2}-n}}\frac{A^2}{\varepsilon_\infty}\frac{1}{x^3} + 0\left(\frac{1}{x^4}\right),$$

$$\tag{19}$$

$$\tau^{(1)} = -\frac{4}{25}\,\mathrm{Pr}^{-1}\left(\frac{2}{5}\,\varepsilon_\infty\right)^n A + 0\left(\frac{1}{x}\right)$$

(subsonic flow).

The subsequent approximations can be found by solving similar equations; as can easily be seen, they do not yield essentially new results.

Apart from the solutions of the equations for the approximations of different orders, (13) and (14) permit one to obtain the zeroth approximation for the Knudsen number

$$\alpha^{(0)} = \alpha_0\sqrt{2\varepsilon_\infty}\left(\frac{w_0}{\sqrt{2\varepsilon_\infty}}\right)^{\frac{2}{3}\left(n-\frac{1}{2}\right)}\frac{1}{x^{\frac{4n}{3}-\frac{5}{3}}}\left[1 + 0\left(\frac{1}{x}\right)\right] \tag{20}$$

(supersonic flow);

$$\alpha^{(0)} = \alpha_0\left(\frac{2}{5}\,\varepsilon_\infty\right)^{n-2}w_0\,\frac{1}{x} + 0\left(\frac{1}{x}\right) \tag{21}$$

(subsonic flow).

Certain conclusions may be drawn from an analysis of the expressions for the first approximation (18) and (19) and for the local Knudsen number (20) and (21):

A) In the case of a subsonic flow the contribution of the first and subsequent terms of the expansion in Re_0^{-1} is small compared with that of the zeroth term; the equations (14) give the basic properties of the solutions of the system (3) for arbitrarily large values of x. The local Knudsen number decreases monotonically; therefore, if a subsonic flow obtains everywhere, there is no possibility of the role of the dissipative terms increasing, nor of the occurrence of a flow with a transition from gas-dynamic to free-molecular conditions.

B) In the case of a supersonic flow the contribution of the first and subsequent terms of the expansion in Re_0^{-1} depends greatly on the parameter n. It follows from (13) and (18) that if $n > 7/4$ the order of the first term in the expansion is higher than that of the zeroth term (the same also holds for the subsequent terms in the expansion); in this case the basic properties of the solutions up to any distances x may be recovered from the zeroth approximation. If $n < 7/4$ the first and subsequent terms in the expansion (10) are of lower order than the zeroth. Thus, the zeroth approximation can no longer give the properties of the solutions of the system (3) at large values of x, since, beginning with some distance, the role of the dissipative terms becomes significant. Under these conditions it is impossible to obtain the asymptotic behavior of the solutions of the system of equations (3) by using an expansion in a series in Re_0^{-1}, since in such an approach the influence of viscosity and thermal conduction on the nature of the solutions cannot be taken into account correctly. The zeroth approximation for the local Knudsen number also reveals that at a certain value of x the dissipative terms begin to play an important role. Moreover, one can easily see from (20) that $\alpha \to \infty$ for $n < 5/4$. However, this does not necessarily imply that a free-molecular asymptotic behavior results. The point is that an increase in the local Knudsen number means, in the first place, that the dissipative terms begin to play an important role; in turn, this may considerably alter the behavior of the solutions for large values of x. Valid conclusions about the asymptotic behavior of the solutions of the system (3) and whether the gas-dynamic approximation is violated can only be obtained if the asymptotic behavior of the solutions of the system (3) for arbitrary values of Re_0^{-1} is known. If analytic means fail to solve this problem, recourse must be had to numerical methods.

3. Besides the case considered in Section 2, the behavior of the solutions of the system (3) can be investigated analytically in a further case:

$$Re_0^{-1} \ll 1, \quad Pe_0^{-1} \gg 1. \tag{22}$$

The inequalities (22) may hold by virtue of $Pr^{-1} \gg 1$, although, in contrast to the case $Re_0^{-1} \ll 1$, and $Pe_0^{-1} \ll 1$, the inequalities (22) cannot be made arbitrarily strong. A solution of the system (3) may be found as an asymptotic series in the small parameter Pe_0

$$\begin{aligned}
w &= w^{(0)} + Pe_0 w^{(1)} + \ldots, \\
\tau &= \tau^{(0)} + Pe_0 \tau^{(1)} + \ldots,
\end{aligned} \tag{23}$$

the functions $w^{(0)}, w^{(1)}, \ldots, \tau^{(0)}, \tau^{(1)}, \ldots$ being determined by the following system of equations:

$$\left.\begin{aligned}
x^2 \tau^{(0)n} \frac{d\tau^{(0)}}{dx} + Pe_0 \varepsilon_\infty &= 0, \\
Re_0^{-1} x^2 \tau^{(0)n} w^{(0)2} \frac{d^2 w^{(0)}}{dx^2} + Re_0^{-1} x^2 \tau^{(0)(n-1)} w^{(0)3} \left(\frac{2\tau^{(0)}}{x} + n\frac{d\tau^{(0)}}{dx} \right)\left(\frac{dw^{(0)}}{dx} - \frac{w^{(0)}}{x} \right) - \\
- (w^{(0)2} - \tau^{(0)}) \frac{dw^{(0)}}{dx} - w^{(0)} \frac{d\tau^{(0)}}{dx} - w^{(0)} \frac{A}{x^2} + 2\frac{w^{(0)}\tau^{(0)}}{x} &= 0.
\end{aligned}\right\} \tag{24}$$

In the zeroth approximation for the energy equation we assume that the energy flux at the initial level is primarily due to molecular heat conduction. The solution of the first equation of the system (24) has the form

$$\tau^{(0)} = \left\{ \frac{(n+1)\,Pe_0\varepsilon_\infty}{x} + [1 - (n+1)\,Pe_0\,\varepsilon_\infty] \right\}^{\frac{1}{n+1}}, \tag{25}$$

and if $\tau^{(0)}(x)$ tends to zero at infinity, (25) reduces to

$$\tau^{(0)} = \frac{[(n+1)\,Pe_0\varepsilon_\infty]^{\frac{1}{n+1}}}{x^{\frac{1}{n+1}}}. \tag{25'}$$

To solve the second equation we use the inequality $Re_0^{-1} \ll 1$ and represent the solution as a series in Re_0^{-1}:

$$w^{(0)} = w^{(0,0)} + Re_0^{-1} w^{(0,1)} + \cdots \tag{26}$$

Substituting (26) into the second equation of (24), we obtain the following differential equation for $w^{(0,0)}$:

$$(w^{(0,0)^2} - \tau^{(0)}) \frac{dw^{(0,0)}}{dx} + w^{(0,0)} \frac{d\tau^{(0)}}{dx} + w^{(0,0)} \frac{A}{x^2} - 2 \frac{w^{(0,0)}\tau^{(0)}}{x} = 0, \tag{27}$$

which, if $\tau^{(0)}$ is determined by (25'), has a saddle-type critical point and admits subsonic and supersonic solutions at great distances.

The question of the uniform approximation of the solution of the system (3) by the functions $w^{(0,0)}$ and $\tau^{(0)}$ can be investigated by means of the local Knudsen number, calculated for the supersonic branch of (27)

$$\alpha^{(0)} = \alpha_0 w^{(0,0)} [(n+1)\,Pe_0\varepsilon_\infty]^{\frac{2n-1}{2(n+1)}} x^{\frac{3}{2(n+1)}}. \tag{28}$$

Since $w^{(0,0)}$ increases monotonically ($\alpha^{(1)} \to \infty$ as $x \to \infty$ for all n), the system of functions $w^{(0,0)}$ and $\tau^{(0)}$ cannot, after a certain distance, serve as an approximation to the solution of the system (3) if the conditions (22) hold; this is because of the importance that viscosity now acquires. Here, as in the case $n < {}^{7}\!/_{4}$, the inequalities Re_0^{-1}, $Pe_0^{-1} \ll 1$ holding, one requires the asymptotic behavior of the solutions of the system (3) for large values of x and arbitrary values of Re_0^{-1} and Pe_0^{-1}.

4. We now consider the problem of a plasma flowing out from the Sun, taking into account the results we have obtained on the properties of a steady-state, spherically symmetric outflow of gas in the gravitational field of a star. The parameters needed in the investigation can be estimated from observational data on the solar corona and the interplanetary plasma. Fairly crude estimates yield

$$Re_0^{-1} \approx 1, \quad Pe_0^{-0} \approx 30 \text{ to } 40,$$

although these may not be the actual values. Therefore, in analyzing the problem we shall consider two cases:

$$Re_0^{-1} \ll 1, \; Pe_0^{-1} \ll 1 \text{ and } Re_0^{-1} \ll 1, \; Pe_0^{-1} \gg 1, \; Re_0^{-1} \approx 1, \; Pe_0^{-1} \gg 1.$$

In the first case the behavior of the hydrodynamic quantities at large distances can be described by the expressions

$$w^{(1)} = \frac{1}{3} (2\varepsilon_\infty)^{1/s} \left(\frac{w_0}{\sqrt{2\varepsilon_\infty}} \right)^{s/s} \frac{1}{x^{7/s}} \left[1 + 0\left(\frac{1}{x} \right) \right],$$

$$\tau^{(1)} = -\frac{14}{15} \varepsilon_\infty \left(\frac{w_0}{\sqrt{2\varepsilon_\infty}} \right)^{s/s} \frac{1}{x^{1/s}} \left[1 + 0\left(\frac{1}{x} \right) \right], \tag{29}$$

obtained from (18) for the case $n = \frac{5}{2}$ (only the supersonic branch has been taken, since the subsonic branch is unstable according to the existing data on the properties of the solar corona and the interplanetary plasma). The local Knudsen number for the supersonic branch is given in the zeroth approximation by the formula

$$\alpha = \alpha_0 \sqrt{2\varepsilon_\infty} \left(\frac{w_0}{\sqrt{2\varepsilon_\infty}} \right)^{4/5} \frac{1}{x^{2/5}} \left[1 + 0 \left(\frac{1}{x} \right) \right]. \tag{30}$$

An investigation of (29) and (30) shows that at large distances the contribution of the dissipative terms is small and the zeroth approximation describes the plasma flow from the sun with an adequate accuracy. In other words, if the flow is such that the inequalities $\mathrm{Re}_0^{-1} \ll 1$ and $\mathrm{Pe}_0^{-1} \ll 1$ are satisfied at the initial level, it is not only impossible for a transition to free-molecular conditions to occur, but also impossible for dissipative processes to exert a significant influence. Certain difficulties arise in interpreting the plasma flow from the sun by solutions of this type: they are connected with the mechanisms responsible for accelerating the plasma to the observed energies (heat sources of great extension are needed) and also with the behavior at large distances, since the interaction of the outflowing plasma with the interplanetary medium has a gas-dynamic nature.

In the second case the behavior of the hydrodynamic quantities at large distances can only be found if the asymptotic behavior of the solutions at large values of x is known for arbitrary Re_0^{-1} and Pe_0^{-1}. If $\mathrm{Re}_0^{-1} \approx 1$ and $\mathrm{Pe}_0^{-1} \gg 1$, this can be seen directly. However, if $\mathrm{Re}_0^{-1} \ll 1$, then, as was shown in the preceding section, the solution behaves in such a way that, beginning at a certain distance, the viscous terms cannot be neglected and the principal question can only be answered if the behavior of the solutions at large distances is known for arbitrary values of Re_0^{-1} and Pe_0^{-1}. Since we have not yet succeeded in finding the asymptotic behavior of the solutions of the system of equations (3) for arbitrary values of Re_0^{-1} and Pe_0^{-1}, only numerical methods can determine whether gas dynamics may be applied and whether a flow can occur with a finite radius of the region of frequency collisions.* In order to carry out a numerical integration we need data on the conditions of the initial level, not only for the determination of the parameters of the problem, but also for the initial values of the velocity and the temperature and their first derivatives. In the integration we took the following values of the parameters of the problem and initial conditions:

$$\mathrm{Re}_0^{-1} = 1.52, \quad \mathrm{Pe}_0^{-1} = 42.53,$$
$$w_0 = 0.17, \quad \tau_0 = 1, \left(\frac{dw}{dx} \right)_0 = 0.44, \left(\frac{d\tau}{dx} \right)_0 = -0.24. \tag{31}$$

The system of equations (3) was reduced to the normal form

$$\frac{dw}{dx} = u,$$

$$\frac{du}{dx} = -\left(\frac{2}{x} + 2.5 \frac{\pi}{x} \right) \left(u - \frac{w}{x} \right) - 0.66 \frac{u}{wx^2\tau^{5/2}} \left(\frac{\tau}{w} - w \right) + 0.66 \frac{\pi}{wx^2\tau^{5/2}} + 3.14 \frac{1}{wx^4\tau^{5/2}} - 1.32 \frac{1}{wx^3\tau^{5/2}},$$

$$\frac{d\tau}{dx} = \pi,$$

$$\frac{d\pi}{dx} = -2.5 \frac{\pi^2}{x} - 2 \frac{\pi}{x} - 0.03 \left(u - \frac{w}{x} \right)^2 - 0.035 \frac{\pi}{x^2\tau^{5/2}} + 0.023 \frac{u}{wx^2\tau^{5/2}} + 0.046 \frac{1}{x^2\tau^{5/2}}, \tag{32}$$

---

*A solution of the problem of numerical methods is necessary even if one does succeed in finding the asymptotic behavior of the solutions for large values of x and in proving analytically that a hydrodynamic model with a finite radius of the region of frequent collisions can exist. The point is that in actually constructing a model one needs to know the distance from the initial level at which the conditions for the applicability of gas dynamics begin to be violated; this can only be found by numerical methods.

TABLE 1

x	w	u	τ	π	μ	α	β
1.00	0.17	0.440	1.00	—0.250			
1.10	0.21	0.430	0.98	—0.212	0.17	0.18	—1.85
1.20	0.26	0.420	0.96	—0.182	0.20	0.22	—1.53
1.30	0.30	0.410	0.94	—0.158	0.24	0.27	—1.26
1.40	0.34	0.390	0.93	—0.138	0.27	0.32	—1.02
1.50	0.37	0.380	0.91	—0.122	0.30	0.37	—0.81
1.60	0.41	0.370	0.90	—0.109	0.33	0.43	—0.63
1.70	0.45	0.350	0.89	—0.097	0.37	0.48	—0.47
1.80	0.48	0.340	0.88	—0.088	0.40	0.54	—0.32
1.90	0.52	0.330	0.873	—0.079	0.43	0.60	—0.18
2.00	0.55	0.320	0.866	—0.072	0.45	0.66	—0.06
2.10	0.58	0.320	0.860	—0.066	0.48	0.72	0.05
2.20	0.61	0.310	0.85	—0.060	0.51	0.78	0.16
2.30	0.64	0.310	0.847	—0.055	0.54	0.85	0.23
2.40	0.67	0.300	0.842	—0.051	0.57	0.92	0.35
2.50	0.70	0.300	0.837	—0.047	0.59	0.90	0.44
2.60	0.73	0.300	0.832	—0.044	0.62	1.06	0.52
2.70	0.76	0.290	0.828	—0.040	0.65	1.13	0.60
2.80	0.79	0.290	0.824	—0.038	0.67	1.21	0.68
2.90	0.82	0.290	0.820	—0.035	0.70	1.28	0.75
3.00	0.85	0.290	0.827	—0.033	0.73	1.36	0.82
3.10	0.88	0.290	0.814	—0.031	0.75	1.45	0.89
3.20	0.91	0.290	0.811	—0.029	0.78	1.53	0.96
3.30	0.94	0.280	0.808	—0.027	0.80	1.62	1.02
3.40	0.97	0.280	0.805	—0.026	0.83	1.70	1.08
3.50	0.99	0.280	0.803	—0.024	0.85	1.79	1.14
3.60	1.02	0.280	0.801	—0.023	0.88	1.89	1.20
3.70	1.05	0.280	0.798	—0.021	0.91	1.98	1.26
3.80	1.08	0.280	0.796	—0.020	0.93	2.08	1.32
3.90	1.11	0.280	0.794	—0.019	0.96	2.18	1.38
4.00	1.13	0.280	0.792	—0.018	0.98	2.28	1.44
4.10	1.16	0.280	0.791	—0.017	1.01	2.39	1.50
4.20	1.19	0.280	0.789	—0.016	1.03	2.49	1.55
4.30	1.22	0.280	0.787	—0.016	1.06	2.60	1.61
4.40	1.25	0.230	0.786	—0.015	1.08	2.71	1.66
4.50	1.27	0.280	0.784	—0.014	1.11	2.83	1.72
4.60	1.31	0.280	0.783	—0.014	1.14	2.95	1.78
4.70	1.33	0.280	0.782	—0.013	1.16	3.06	1.83
4.80	1.36	0.280	0.780	—0.012	1.19	3.19	1.89
4.90	1.39	0.290	0.779	—0.012	1.21	3.31	1.95
5.00	1.42	0.290	0.778	—0.011	1.24	3.44	2.00
5.10	1.45	0.290	0.777	—0.011	1.26	3.57	2.06
5.20	1.48	0.290	0.776	—0.010	1.29	3.70	2.12
5.30	1.51	0.290	0.774	—0.010	1.32	3.83	2.17
5.40	1.53	0.290	0.774	—0.010	1.34	3.97	2.23
5.50	1.56	0.290	0.773	—0.009	1.37	4.11	2.29
5.60	1.59	0.290	0.772	—0.009	1.39	4.25	2.35
5.70	1.62	0.290	0.771	—0.009	1.42	4.39	2.40
5.80	1.65	0.290	0.770	—0.008	1.45	4.54	2.47
5.90	1.68	0.290	0.769	—0.008	1.47	4.69	2.53
6.00	1.71	0.291	0.769	—0.008	1.50	4.84	2.59
6.50	1.85	0.291	0.765	—0.006	1.62	5.64	2.90
7.00	2.00	0.292	0.762	—0.005	1.76	6.49	3.22

TABLE 1 (concluded)

x	w	u	τ	π	μ	α	β
7.50	2.14	0.293	0.760	—0.005	1.89	7.42	3.56
8.00	2.29	0.295	0.758	—0.004	2.02	8.41	3.92
8.50	2.44	0.296	0.756	—0.004	2.16	9.47	4.30
9.00	2.59	0.297	0.754	—0.003	2.29	10.59	4.70
9.50	2.73	0.297	0.753	—0.003	2.43	11.77	5.12
10.00	2.88	0.298	0.751	—0.002	2.56	13.03	5.55
10.50	3.03	0.299	0.750	—0.002	2.70	14.34	6.02
11.00	3.18	0.299	0.749	—0.002	2.83	15.72	6.50
11.50	3.33	0.300	0.748	—0.002	2.97	17.17	7.01
12.00	3.48	0.300	0.748	—0.002	3.10	18.68	7.53
12.50	3.63	0.300	0.747	—0.001	3.24	20.26	8.08
13.00	3.78	0.301	0.746	—0.001	3.37	21.90	8.65
13.50	3.93	0.301	0.746	—0.001	3.51	23.61	9.24
14.00	4.08	0.301	0.745	—0.001	3.64	25.39	9.86
14.50	4.23	0.302	0.745	—0.001	3.78	27.23	10.50
15.00	4.38	0.302	0.744	—0.001	3.91	29.03	11.16
15.50	4.54	0.302	0.744	—0.001	4.05	31.10	11.84
16.00	4.69	0.302	0.743	—0.001	4.19	33.14	12.54
16.50	4.84	0.302	0.743	—0.001	4.32	35.24	13.27
17.00	4.99	0.303	0.743	—0.001	4.46	37.41	14.02
17.50	5.14	0.303	0.742	—0.001	4.59	39.65	14.80
18.00	5.29	0.303	0.742	—0.001	4.73	41.95	15 59
18.50	5.44	0.303	0.742	—0.001	4.87	44.31	16.41
19.00	5.60	0.303	0.741	—0.001	5.00	46.74	17.25
19.50	5.74	0.303	0.741	—0.001	5.14	49.24	18.12
20.00	5.90	0.303	0.741	0	5.27	51.80	19.01
20.50	6.05	0.303	0.741	0	5.41	54.43	19.92
21.00	6.20	0.303	0.741	0	5.55	57.13	20.85
21.50	6.35	0.304	0.740	0	5.09	59.89	21.81

and by means of integration on a computer the following results were obtained: the values of w, u, τ, and π as functions of x; the local Mach number $\mu = 0.67w/\tau^{1/2}$; the dimensionless flux of hydrodynamic energy per particle $\beta = w^2/2 + \tfrac{5}{2}\tau - 4.75/x$; and the local Knudsen number $\alpha = \lambda/r = 0.80w\tau^2 x$. The results of the numerical integration for the above version of the problem on the M-20 computer are given in Table 1; they show that the conditions for the applicability of the gas-dynamic approximation are violated at distances beginning with values of x of the order of 2.5–3$R_\odot$. Thus, under the given assumptions about the conditions at the initial level one must use a hydrodynamic model with a finite-radius region of frequent collisions to describe the system consisting of the solar corona and the interplanetary plasma.

5. Finally, it is of interest to mention a curious circumstance. In problems on the steady-state flow of a gas from stars it is found that conditions arise under which the criteria for the applicability of gas dynamics are violated. However, before free-molecular conditions occur, there must be a region of flow in which dissipative processes play a significant role, even if they were of practically no importance at all at the initial level. This is connected with an increase in the mean-free path and the transport coefficients (the thermal conductivity and viscosity) to values at which the dissipative terms become important, even though the values of the gradients of the hydrodynamic quantities are relatively small. It is precisely this phenomenon that explains why, as we have found, it is impossible for there to be a transition from gas dynamics to free-molecular conditions in the transition layer in the corona (the possibility we investigated was of a transition from nonviscous and non-heat-conducting gas dynamics to free-molecular conditions) [4]. It seems to us that this circumstance must be taken into account in an investigation of the flow of a gas into space if the parameters indicate that free-molecular conditions could arise. *

It is a pleasant duty to thank I. M. Dagkesamanskii for carrying out the numerical integration of the system of equations (32) on the M-20 computer.

LITERATURE CITED

1. K. J. Gringans, Space Res., 2:539 (1962); M. Nengetaner and C. W. Snyder, 138:1095 (1962).
2. J. H. Chisholm and J. C. James, Astrophys. J., 140:377 (1964).
3. E. N. Parker, Interplanetary Dynamical Processes (Interscience Monographs and Texts in Physics and Astronomy, Vol. 8), New York (1963).
4. M. V. Konyukov, in: Wideband Cruciform Radio Telescope Research (D. V. Skobel'tsyn, ed.), Consultants Bureau, New York (1969), p. 115.
5. J. Chamberlain, Astrophys. J., 131:47 (1960); L. M. Noble and F. L. Scearf, Astrophys. J., 138:1169 (1963).
6. Wai Shih-i, Magnetohydrodynamics and Plasma Dynamics [Russian translation], IL, Moscow (1964).
7. R. Deich, Cosmic Gas Dynamics [Russian translation], IL, Moscow (1964).
8. Ch'ien Hsüeh-sen, Problemy Mekhaniki, No. 11, p. 7 (1953).
9. G. Gred, Various Questions of the Kinetic Theory of Gases, Izd. Mir, Moscow (1965).

---

*It is noteworthy that the course of the relaxation of the free-molecular flow to a hydrodynamical flow is as follows: free-molecular flow → nonviscous gas dynamics → viscous gas dynamics [9].

GAIN STABILIZATION OF TRANSISTORIZED SELECTIVE
LF AMPLIFIERS IN RADIOMETRIC RECEIVERS

V. N. Brezgunov and V. A. Udal'tsov

Vacuum-tube low-frequency tuned amplifiers are used in radiometric receiving devices, a typical circuit being that in [1], p. 82. The selectivity is provided by double-T networks in the negative feedback circuit. Such amplifiers have been found to be very satisfactory.

However, multichannel and transportable equipment requires higher reliability and reduced size, so transistorized circuits have become necessary, and this has raised the problem of gain stabilization. It has been stated [2] that it is better to use automatic control circuits rather than thermostatic systems in transistor amplifiers.

We have made a transistorized lf amplifier with gain stabilization via an automatic gain control (AGC), which consists of a two-frequency selective system (Fig. 1), the two frequencies being widely separated. The AGC signal is transformed from a dc-reference signal into an ac one whose frequency is substantially different from that of the working signal. The two signals pass through the amplifying stages and are separated by frequency at the output; the AGC signal is then converted to dc and used to control the gain, while the working signal is passed to a phase-sensitive detector. The AGC signal alters with the actual gain and adjusts that gain to the set value within the accuracy of the control.

This method cannot be based on double-T RC filters, since the active components of the selective circuits must be the same. A resonant system employing ferrites was therefore used, which is no less selective than the double T (Fig. 2). The lf amplifier with AGC is tuned to two different frequencies by two resonant systems. The selectivity at the working frequency is provided by the tuned systems K_1 and K_2 at the input and output together with a phase-sensitive detector. Circuit K_3 provides the selectivity at the AGC frequency. The amplifier is made aperiodic (apart from the last stage) in order to transmit both frequencies equally. The load in the last stage is K_2 and K_3 in series, which work respectively into the phase detector and AGC detector.

This system provides quite adequate isolation between the AGC and working signals. The AGC signal passes through the entire amplifier and carries information about the gain at its frequency. The gain varies in the same way at all frequencies in response to the control signal, so the AGC signal can be used for automatic gain control at both frequencies.

The usual series system (T_1-T_3) [2, 3] is used in gain control, in which T_2 and T_3 are controlled elements that act as the dc load for T_1. The control voltage adjusts the current through T_1 and hence that through T_2 and T_3, which adjusts the gain of the latter.

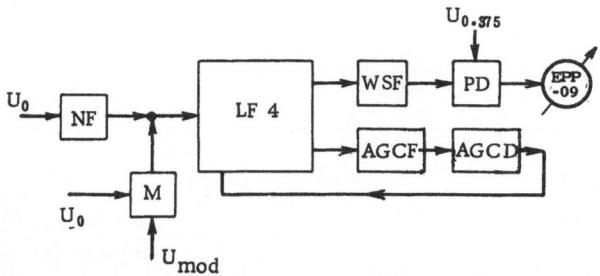

Fig. 1. Block diagram of gain stabilization in an lf amplifier. U_c, 375-Hz working signal; U_0, dc reference voltage of the AGC system; U_{mod}, 10-kHz modulating voltage; $U_{0.375}$, reference voltage for the phase-sensitive detector; NF, noise filter; M, modulator; WSF, working signal filter; AGCF, AGC filter; AGCD, AGC Detector; PD, phase-sensitive detector.

Allowance must be made for the strength and spectrum of the noise at the amplifier input in using the device in a radiometric receiver. The low-frequency component in the current fluctuations is as follows for a linear detector, while the dc component is

$$\overline{I_l} = 0.2S \sqrt{u_s^2},$$

$$I_= = 0.4S \sqrt{u_s^2},$$

where S is the slope of the detector, and $\overline{u_s^2}$ is the mean square of the input noise.

For a square-law detector,

$$\overline{I_l} = \overline{I_=} = \beta \sqrt{u_s^2},$$

where β is the coefficient in $I = \beta U^2$. The ratio of the ac and dc current components is 0.5 for a linear detector and 1 for a square-law one.

If the detector output due to receiver noise is 1 V, the equivalent voltage of the low-frequency component of the noise spectrum is 0.5 V for a linear detector and 1 V for a quadratic one. Such a noise level would overload a transistor amplifier, so it is necessary to use a passive narrow-band filter tuned to the working frequency at the input.

If the receiver has a rectangular characteristic with a passband ΔF = 100 kHz, while the filter for the working frequency has Δf = 30 Hz, the noise voltage at the filter output is reduced roughly by a factor of $2\Delta f/\Delta F = 6 \cdot 10^{-4}$, which means 0.3 mV output for a linear detector and 0.6 mV for a square-law one. The AGC signal is injected after the filter, with appropriate decoupling between the filter and the AGC modulator.

The working frequency in the present case was 375 Hz, which was chosen on the basis of the radio astronomy system. The necessary selectivity is provided by F-400, 600, and 1000 nickel—zinc ferrites in the conventional core shape, these having a natural $Q \geq 20$. Such Q can be provided down to about 200 Hz, but Q decreases at lower frequencies on account of resistive loss in the copper. There is no upper frequency limit to the use of tuned systems with such Q, though a different type of core is desirable above 1-10 kHz.

We used Sh20 × 28 F-600 ferrite in circuit K_2 (375 Hz). A small bandwidth was provided by making the coupling to the collector of T_5 weak, which gave an equivalent Q of 11. The input circuit at 375 Hz consisted of Sh7 × 7 F-1000 ferrite. The measured equivalent Q in the tapped circuit was 12.

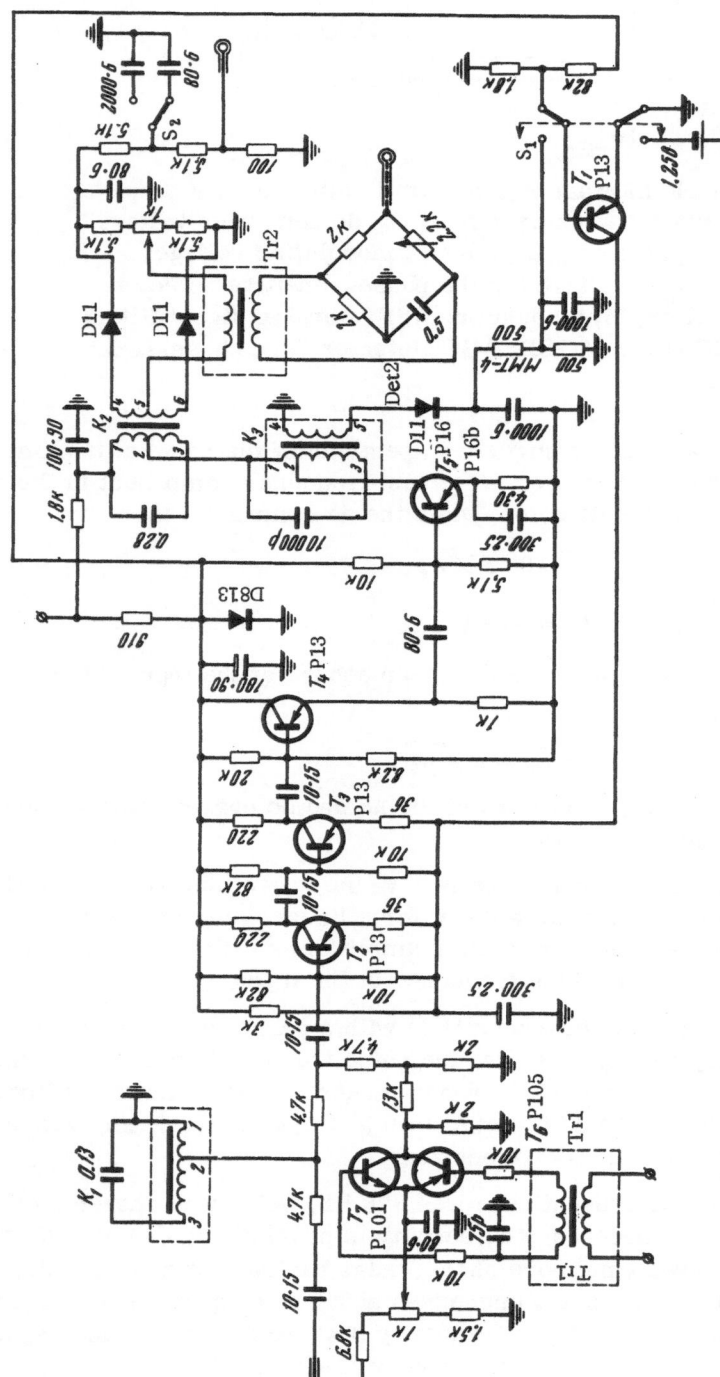

Fig. 2. Theoretical circuit of a selective 1f amplifier with gain stabilization and phase detection. S_1 selects the time constant, S_2 switches the AGC. Windings of coils and transformers: K_1) $n_{13} = 1300$, $n_{12} = 260$, 0.2-mm wire; K_2) $n_{13} = 910$, $n_{12} = 118$, 0.55-mm wire, $n_{45} = 118$, 0.35-mm wire; K_3) $n_{13} = 778$, $n_{12} = 124$, 0.07×7 litz wire, $n_{45} = 62$, 0.16-mm wire; Tr1) 7×7 F-1000 core, 2-sectional windings of 1140 turns, 0.12-mm wire; Tr2) the same, but 2500 turns of 0.1-mm wire.

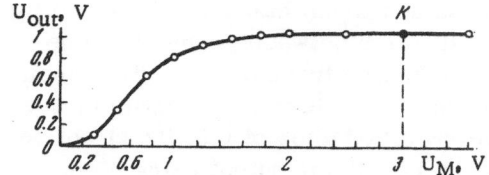

Fig. 3. Modulator output U_{out} as a function of modulating voltage, with the reference voltage at the modulator input kept constant. K is the working point of the modulator.

Good decoupling between the two frequencies at the output is essential. The level of decoupling is governed primarily by the frequency difference. Taking 1% as the maximum permissible coupling, we have $U/U_{res} = 10^{-2}$ for $Q_{eq} = 15$ at 2.5 kHz, via the formula

$$f = f_{res} \frac{1}{Q_{eq} \frac{U}{U_{res}}},$$

which follows from the usual equation for the resonance curve for $Q > 1$ and $f > f_{res}$. The AGC frequency was chosen as 10 kHz, while K_3 was made on an SB-5 screened core with weak coupling to the collector of T_5 and to the AGC detector. The measured equivalent Q was 18.

Good linearity in the amplitude response at both frequencies is also necessary in order to obtain minimal crosstalk. Nonlinearity causes the AGC signal to be modulated by the working signal and also causes distortion of the working signal by an amount dependent on the level of the AGC signal. A decoupling factor of 100 can be provided if the deviation from linearity does not exceed 5%.*

The AGC reference voltage may be strictly constant or may vary in accordance with some law. Modulator T_6 and T_7 transforms the reference voltage to ac and employs a switching system [5], which has the advantage of relatively low level of parasitic signal at the modulation frequency. The modulator has input and output resistances of 2 kΩ, which gives a parasitic voltage at the modulation frequency of about 1 mV for 3-V control voltage.† This is at-

TABLE 1

$-U_b$, V	I_{C_1}, mA	I_{C_2}, mA	I_Σ, mA	$-U_C$, V	R_{in}, Ω
0.13	0.15	0.16	1.40	11.25	900
0.14	0.18	0.18	1.45	11.05	880
0.15	0.28	0.25	1.80	10.80	735
0.16	0.32	0.34	2.00	10.50	680
0.17	0.84	0.62	3.10	9.45	530
0.18	1.10	0.78	3.80	8.80	450
0.19	1.60	1.04	4.70	8.00	400
0.20	2.00	1.29	5.50	7.10	—
0.21	2.90	1.79	7.40	5.70	—
0.22	3.60	2.13	8.80	4.50	—
0.23	4.50	2.55	10.30	3.35	—
0.24	4.90	2.80	11.30	2.75	—
0.25	5.70	3.20	12.90	1.70	—
0.26	6.40	3.50	14.00	0.70	—
0.27	6.80	3.70	14.80	0.35	—

---

*There is no effect here from nonlinearity in the amplitude characteristic due to the properties of the ferrite in K_2.

†The effective parasitic voltage is actually about 3 mV on account of harmonics of the modulation frequency, but these can be neglected.

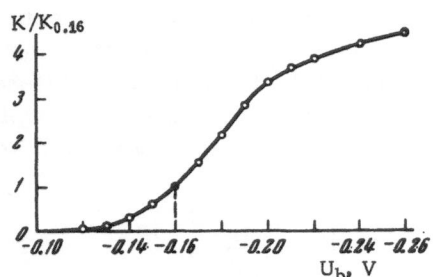

Fig. 4. Gain of the lf amplifier as a function of control voltage, as normalized to the gain at 0.16 V.

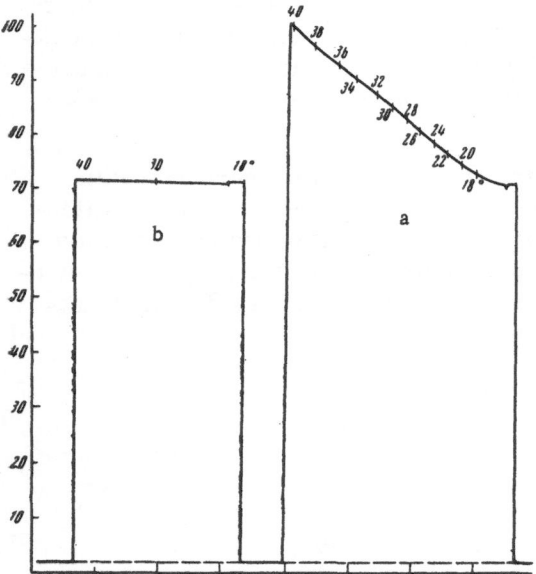

Fig. 5. Recordings of the working signal after passage through the lf amplifier as the temperature varies. Amplifier input kept constant. a) AGC inoperative; b) AGC operating. The numbers on the curves are the air temperatures. The horizontal axis is divided at 10-min intervals.

tained via the additional balancing provided by the 75 pF capacitor between ground and one end of the modulator transformer. The modulator operates on a reference potential of 200 mV with a transfer factor of 0.9, the effective voltage at the modulator output being 65 mV, and the parasitic signal being 1.8%. The AGC signal from the modulator output determines the gain stability, so the output voltage must be independent of time, temperature, and amplitude of the modulating voltage. Time and temperature tests on the modulator at 20-50°C showed the transfer factor and the parasitic signal to be constant to better than 0.2%. Figure 3 shows that the dependence of the transfer factor on the modulating voltage was not so satisfactory; the transfer factor alters by 2% when the modulating voltage deviates 30% from 3 V, so the modulating voltage must be stabilized to 1-3%, which requires special measures.

The requirement of high AGC stability also imposes a requirement for stability in the modulating frequency. Circuit K_3 had an equivalent Q of 11; a permissible change of 0.1% in output voltage due to frequency change implies a frequency shift of $2 \cdot 10^{-3}$, which can occur with a vacuum-tube oscillator fitted with a thermistor.

The amplifier itself employs four transistors. T_2 and T_3 are adjustable in gain, while T_4 is an emitter follower that matches the control stages to the output. T_5 has an input impedance of about 500 Ω and a gain for the working signal of about 100. The high gain in the last stage is necessary in order to provide a high upper limit in linearity for the output voltage. The last stage is fed from a 30-V source via appropriate filters in order to provide decoupling from the other stages.

Table 1 gives the parameters of the transistors in the control stages for various control voltages U_b at the base of the control transistor, while the I_c are the collector currents of the controlled transistors, I_Σ is the total current (collector current of the control transistor), U_c is the collector potential of the control transistor, and R_{in} is the input impedance of the control transistor, as measured with an L2-2 instrument.

The nominal working point is with $U_b = -0.16$ V. The characteristic of Fig. 4 shows that the dynamic control range is 10^3 for U_b from −0.1 to −0.26 V. The control slope [2] is $S_c = \Delta K / K \Delta U_b$, with $S_c = 40$ V^{-1} at $U_b = -0.16$ V.

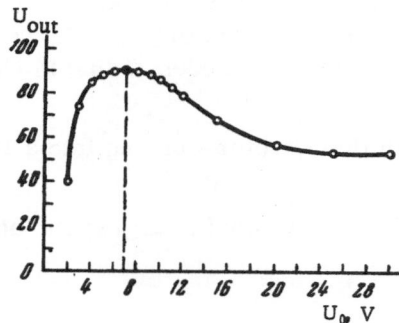

Fig. 6. Output voltage of phase detector as a function of reference voltage at constant signal voltage (output in scale divisions of EPP-09 recorder).

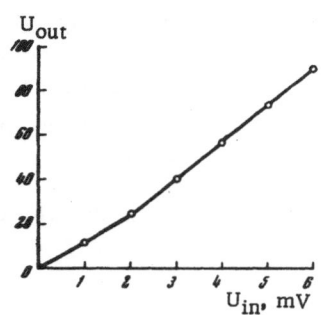

Fig. 7. Amplitude response of selective amplifier recorded on working signal with AGC operative. U_{out} given in scale divisions of EPP-09 recorder.

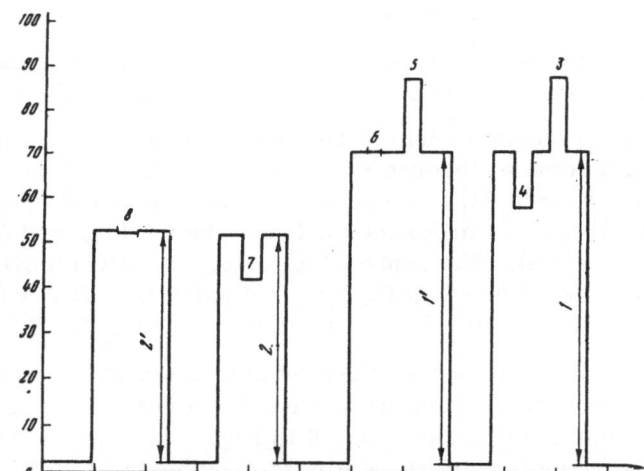

Fig. 8. Recordings of 1,1',3-6) working signal at amplifier output; 2,2',7,8) AGC control signal with AGC on and off; 1,1') working signal with AGC off and on, respectively; 3,5) change in working signal with AGC off and on; 4,6) change in gain with AGC off and on; 2,2') AGC signal with AGC off and on; 7,8) change in gain with AGC off and on.

TABLE 2

Diode type	Temperature range, °C	Change in transfer factor, %
D11	25–50	1.0
D226	26–50	4.3
D9K	20–50	2.4
D2A	20–50	1.4

The AGC substantially improves the stability in response to temperature change over the range 16–40°C (Fig. 5), where the gain without AGC increases by about 40%, as against not more than 0.7% with the AGC operating. The AGC circuit includes an MMT-4 resistor, which compensates the temperature variation in the characteristic of the control transistor.

The phase-sensitive detector employs D-11 diodes. The reference voltage is supplied through a bridge phase shifter, which gives a phase range of 0–140°C at fixed refer-

ence amplitude. Figure 6 shows that the optimal reference voltage is 7 V and that ±30% variations in that voltage produce only 0.2% change in the transfer factor. Stability to 0.1% requires
stabilization of the reference voltage. Measurements on four types of diodes (Table 2) show
that type D11 is the best as regards temperature stability.

All the results reported in Table 2 were obtained with the detector working into a 10-kΩ
load.

The low-pass filter in PD was of T type and provided roughly equal charging and discharg-
ing time-constants.

Circuit Parameters

The passband at the working frequency was 18 Hz at the 0.7 level and was close to optimal,
being sufficiently narrow to protect the amplifier and phase detector from overload while wide
enough to give an acceptable slope in the phase characteristic and acceptable instability in the
transfer factor on account of frequency change in the working signal.

The overall gain (measured as the ratio of the dc output of PD to the effective voltage at
the input) was 100, the gain of the amplifier itself being 10^3 but the input circuit for the working
signal giving an attenuation of about a factor 10.

Figure 7 shows the amplitude characteristic at the working frequency. The nonlinearity
at low signal levels is due to the properties of the ferrite in circuit K_2.

Figure 8 shows signal recordings with steps in the signal and gain in order to evaluate
the control. These indicate some difference in the response steps of the AGC system to stepwise change in the gain as regards AGC signal 8 and working signal 6, which is due to parasitic effects related to the frequency difference between the working and AGC signals. This effect is small and can be neglected. The control factor for the AGC signal in the nominal mode
is 25, as determined from steps 7 and 8 in Fig. 8. The calculated degeneration factor is [2]:
$K_c = S_c U_{out}$. This implies $K_c = 40$ for $U_{out} = 1$ V and $S_c = 40$ units/Y.

The actual degeneration factor is less than the calculated value, which is due to the negative feedback with respect to base current and the low input resistance of the control transistor.
The degeneration factor indicated by steps 8 and 6 in Fig. 8 exceeds the 25 obtained for the AGC
signal, which is due to the parasitic signal mentioned previously.

The gain drift over 24 h from the time of switching on with the AGC operative was not
more than 0.5%. The sources for the working signal, phase reference, and AGC modulation signal were taken from GZ-33 oscillators without special measures for voltage stabilization. The
temperature varied over the range 18-20°C. The corresponding gain change in 24 h with the
AGC inoperative was 40%.

LITERATURE CITED

1. A. D. Kuz'min and A. E. Samoilovich, Radio Astronomy Methods of Measuring Antenna
   Parameters [in Russian], Izd-vo Sovetsko Radio (1964).
2. V. N. Brezgunov and V. A. Udal'tsov, in: Wideband Cruciform Radio Telescope Research
   (D. V. Skobel'tsyn, ed.), Consultants Bureau, New York (1969), p. 96.
3. Semiconductor Devices and Applications (Ya. A. Fedotov, ed.), [in Russian], No. 9, p. 224
   (1963).
4. V. I. Banimovich, Fluctuation Processes in Radio Receiving Devices [in Russian], Izd-vo
   Sovetskoe Radio (1951).
5. A. S. Krys'ko, Pribory i Tekhn. Eksperim., No. 6, p. 113 (1965).

INVESTIGATION OF RANDOM-VARIABLE MOMENTS OF WIND LOADING ACTING ON A ROTATING ANTENNA

V. P. Nazarov, V. V. Dubarenko, D. G. Stepanov, and A. I. Ukhov

1. Statement of Problem

Experience gained in the planning and operation of powerful radio telescopes shows that disturbances related to wind loads are the main factor impeding accurate tracking of cosmic objects. The wind acts on the guidance system via the antenna, which constitutes an aerodynamic filter transforming the energy of the wind. The wind speed and the moment on the control axle of the tracking drive are associated [1] by the relation

$$M(t) = a(t)v^2(t), \tag{1}$$

where $M(t)$ is the instantaneous value of the moment due to the wind on the control axle, $a(t)$ is the aerodynamic-moment coefficient, and $v(t)$ is the instantaneous value of wind speed.

When a radio telescope tracks a cosmic object the antenna profile changes relative to the wind velocity vector. In connection with this the aerodynamic-moment coefficient also changes, the process of its change being random in the general case. Accordingly the moment $M(t)$ is a random function of time; the determination of its statistical nature is the basic task of this article.

In antenna installations intended for tracking cosmic objects there are two suspension systems, equatorial and horizontal. Here we will consider only installations with a horizontal suspension system tracking an object with respect to its angle of elevation and azimuth.

2. Statistical Characteristics of Wind

The wind never remains constant either in magnitude or in direction. A change in direction can be attributed to the random character of the change of the aerodynamic coefficient. The instantaneous value of the wind speed can be represented [1] by the following expression:

$$v(t) = v_0 + v_1(t), \tag{2}$$

where v_0 is the average value of wind speed, $v_1(t)$ is the instantaneous deviation of the wind speed from the average value. Then the moment is defined as

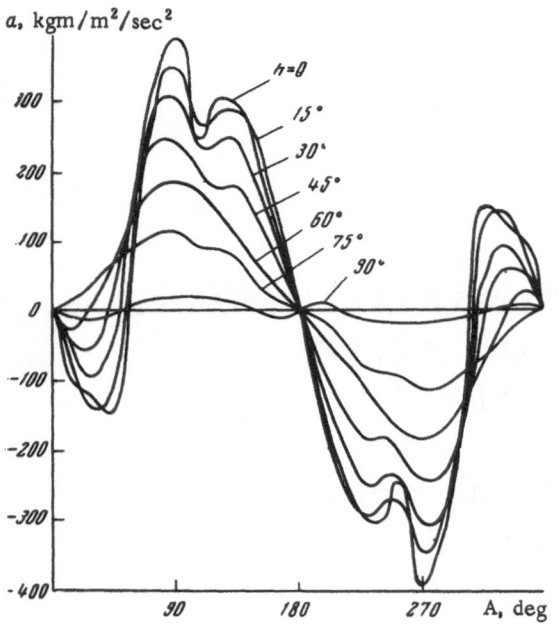

Fig. 1. Aerodynamic moment coefficient based on wind-tunnel results.

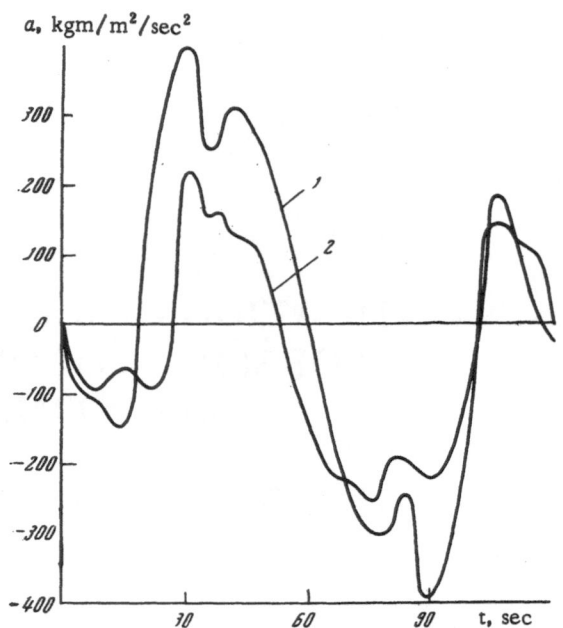

Fig. 2. Time dependence of aerodynamic moment coefficient. 1) Based on wind-tunnel results; 2) based on results of full-scale tests (ω_A = 3 deg/sec, n = 0°).

$$M(t) = a(t) \; [v_0^2 + 2v_0 v_1(t) + v_1^2(t)]. \tag{3}$$

Adding and subtracting in the brackets of expression (3) the quantity $\langle v_1^2(t) \rangle$ of the mean square of the wind speed fluctuations, we obtain

$$M(t) = a(t) \; \{[v_0^2 + \langle v_1^2(t) \rangle] + 2v_0 v_1(t) + [v_1^2(t) - \langle v_1^2(t) \rangle]\}. \tag{4}$$

The term $[v_1^2(t) - \langle v_1^2(t) \rangle]$ in expression (4) can be disregarded by virtue of its smallness. Then the fluctuations of the moment due to the wind can represented approximately as

$$M(t) \approx a(t) \; [v_0^2 + \langle v_1^2(t) \rangle + 2v_0 v_1(t)]. \tag{5}$$

If we denote $v_0^2 + \langle v_1^2(t) \rangle = \langle v^2 \rangle$ the mean square of the wind speed, then

$$M(t) = a(t)[\langle v^2 \rangle + 2v_0 v_1(t)]. \tag{6}$$

We see from expression (6) that the process of fluctuations of the moment on the control axle is a combination of two random processes — fluctuations of wind speed and change of the aerodynamic coefficient. The statistical properties of the wind speed fluctuations have been studied well and their characteristics are presented widely in aerodynamic and other literature.

Thus, on the basis of (1)–(3), the correlation function of wind speed fluctuations can be represented approximately by an exponential curve of the form

$$R_{v_1 v_1}^{(\tau)} = \langle v_1^2(t) \rangle \, e^{-\nu(\tau)}. \tag{7}$$

The values of the constants $\langle v_1^2(t) \rangle$ and ν entering expression (7) are determined by the nature of the wind and depend substantially on wind speed, topography, and the observation interval, which lead to low-frequency oscillations [3].

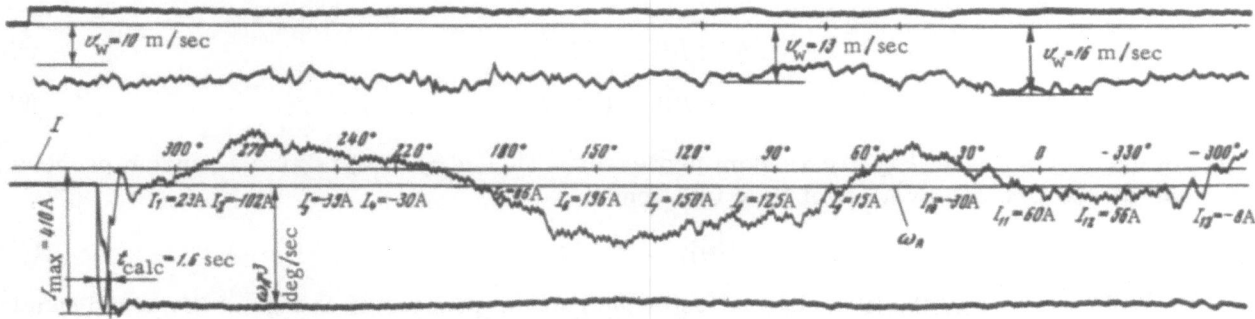

Fig. 3. Recording of the armature current of the control electric motor during movement of the antenna azimuthally (n = 0).

3. Statistical Characteristics of the Aerodynamic Moment Coefficient

The values of the aerodynamic-moment coefficient are determined by the position of the antenna relative to the wind direction. In view of the complexity of the aerodynamic form of the antenna the calculation of coefficient a presents considerable difficulties. In connection with this, the dependence of the aerodynamic-moment coefficient on the azimuth and angle of elevation relative to the wind velocity vector was determined experimentally — by placing a model radio telescope in a wind tunnel [4].

The model was tested at different positions corresponding to different values of azimuth A from 0 to 180° and angle of elevation h from 0 to 90°. The indicated range of angles A and h covers all positions which the antenna can occupy under real conditions, since it has a vertical plane of symmetry. Thus, the position of the antenna with respect to the flow corresponding to values of angles A = 30° and h = 60° is identical to the position at which angles A and h have values of 330 and 60°, respectively. In this case the signs of certain aerodynamic coefficients can be opposite.

Figure 1 shows the family of curves of the change of the aerodynamic-moment coefficient as a function of the angle of rotation with respect to the azimuth A at fixed angles of elevation h obtained as a result of wind-tunnel tests of a model radio telescope with a 32-m parabolic antenna.

Figure 2 shows, for comparison, the curves of the aerodynamic-moment coefficient plotted from the results of wind-tunnel and full-scale tests of a radio telescope.

In the full-scale tests, the aerodynamic-moment coefficient was determined by recording the armature current of the control electric motor of the azimuth drive during movement of the antenna azimuthally at a speed of 3 deg/sec from 0 to 360° and angle of elevation h = 0°. One of the recordings of the armature current is shown in Fig. 3.

When calculating the coefficient a (curve 2, Fig. 2) we disregarded losses due to friction in the reducing gear and changes of wind direction during movement azimuthally, but nonetheless the character of change of curves 1 and 2 (Fig. 2) coincides. We associate the differences of curves 1 and 2 with the different conditions of setting up the experiments in the wind-tunnel and full-scale tests.

We will examine the method of obtaining the statistical characteristics of the aerodynamic coefficient by way of an analysis of the data of the wind-tunnel test on the model radio telescope with a 32-m parabolic antenna. The aerodynamic coefficient, more truly its mathematical model, has a number of special features:

1) $a = f(A_1h)$, but

$$A = f(t) = \{^k A(t)\}, \tag{8}$$

$$h = f(t) = \{^k h\,(t)\}, \tag{9}$$

i.e., in the general case they are random processes. Then $a = f(A_1h)$ is also a random process and can be represented by a set of time functions

$$\{^k a(t)\}, 0 < t < {}^k T, \tag{10}$$

where ${}^k T$ is the time of the k-th realization of the random process of change of the aerodynamic-moment coefficient. The index k is needed to distinguish one realization from another.

2) At the start of movement the position of the antenna azimuthally is equiprobable with respect to the direction of the wind velocity vector, and the position with respect to the angle of elevation $h = 0°$. It is considered that at the start of reception ($t = 0$) the object appears from behind the horizon.

3) The limited power of the guidance drive leads to limitation of the maximum velocities and accelerations:

$$\begin{aligned}
\omega_A(t) &\leqslant \Omega_A, \\
\omega_h(t) &\leqslant \Omega_h, \\
\varepsilon_A(t) &\leqslant \varrho_A, \\
\varepsilon_h(t) &\leqslant \varepsilon_h,
\end{aligned} \tag{11}$$

where $\omega_A(t)$, $\omega_h(t)$, $\varepsilon_A(t)$, and $\varepsilon_h(t)$ are the running values of velocities and accelerations of the antenna with respect to the azimuth and angle of elevation, respectively; and Ω_A, Ω_h, ε_A, and ε_h are the peak values of these velocities and accelerations.

In addition, the velocity characteristics of the antenna are limited by the time of line-of-sign reception T, with respect to the object being observed. Processes $\{^k \omega_A(t)\}$ and $\{^k \omega_h(t)\}$ are random, and for their characterization we must know the mathematical expectation and the distribution law relative to the mathematical expectation or have a set of realizations.

Limitation of the velocities and accelerations leads to the appearance of "blind spots" for telescopes with horizontal suspension systems and consequently to interruptions in communication [5]. Henceforth we will consider continuous communication to be a necessary condition, i.e., movement of an artificial earth satellite (AES) in a spot "blind" for a telescope will not be considered.

4) Consider the uses of radio telescopes: reception of radio signals from natural cosmic objects, and from artificial earth satellites (AES).

In the first case the work of the radio telescope is characterized by slow (up to several minutes of angle per second) rates of change of the azimuth and angle of elevation of the antenna [5]. When considering the dynamics of the effect of wind on the guidance system [1], we can take the aerodynamic-moment coefficient to be a constant.

During operation of a radio telescope with respect to AES, which have orbital altitudes of several hundred kilometers above the earth's surface, the speeds of the antenna with respect to azimuth and angle of elevation reach several degrees per second. This leads to appreciable rates of change of the aerodynamic profile of the antenna, which complicates the problem of determining the statistical characteristics of the change of the aerodynamic-moment coefficient. To solve it we need special investigations of the trajectories of cosmic objects and determination of their theoretico-probabilistic characteristics. However, this is beyond the scope of the present article.

It is obvious that the solution of the problem of determining the statistical properties of the aerodynamic-moment coefficient in the most general case when any trajectory being tracked is equiprobable is not expedient. Apparently, it is necessary to limit oneself to one trajectory or group of trajectories.

We will examine the method of determining the correlation function of a change of the aerodynamic coefficient when tracking any trajectory of a given group. For simplicity of the arguments, we will take trajectories with circular orbits, for which the change of azimuth and angle of elevation, disregarding the angular velocity of the earth's rotation, have the form (see Appendix):

$$A(t) = \tan^{-1}\left[\frac{\omega_\alpha}{\Omega_A}\tan(\omega_\alpha t + \alpha_0)\right],$$

$$h(t) = \tan^{-1}\frac{\sqrt{1 - \dfrac{\omega_\alpha}{\Omega_A}}\sin(\omega_\alpha t + \alpha_0) - \dfrac{R_E}{R_E + H}}{\sqrt{\cos^2(\omega_\alpha t + \alpha_0) + \left(\dfrac{\omega_\alpha}{\Omega_A}\right)^2\sin(\omega_\alpha t + \alpha_0)}},$$

$$(12)$$

where $\omega_\alpha = \dfrac{\sqrt{g(R_E + H)}}{R_E + H} = \sqrt{\dfrac{g}{R_E + H}}$ is the angular velocity of rotation of a satellite around the earth; R_E is the earth's radius; H is the orbital altitude above the earth's surface.

We limit the group of orbits of the AES to the range

$$H_1 < H_\xi < H_u. \tag{13}$$

If now we consider any trajectory to be equiprobable with simultaneous fulfillment of conditions (11)-(13), obviously we obtain a set of time functions

$$\{^\xi A(t)\} \text{ and } \{^\xi h(t)\}, \quad 0 < t < {}^\xi T. \tag{14}$$

These sets characterize the movement of the antenna. Here ξ is the number of the trajectory which takes all values of some interval of the numerical axis from 1 to u, where u is the number of trajectories considered; ${}^\xi T$ is the length of time of the communication (the time of direct visibility of the AES from the observation point). In other words, we have two finite sets of analytically given functions, whereby each pair ${}^\xi A(t)$ and ${}^\xi h(t)$ characterizes the ξ-th trajectory of the AES. For all trajectories a certain combination of the azimuth and angle of elevation corresponds to a certain time, and a certain value of the aerodynamic-coefficient corresponds to these angles from the family of characteristic curves $a(A, h)$ (see Fig. 1).

Dividing the period of communication ${}^\xi T$ into equal time intervals, we find the values of the coefficient a corresponding to the ends of these time intervals, taking any point on the A axis as the origin. Then having changed the origin of A for the same times, we obtain the other values of the coefficient a. The curves of the change $a(t)$ will be equal for different wind directions at the start of movement. The wind direction will be assumed to be equiprobable for any azimuth at the start of movement.

If we take several points of the origin of A, we can regard each curve ${}^\xi a(t)$ as the realization of the random process of change of the aerodynamic coefficient represented by a set of realizations corresponding to different azimuthal wind directions when tracking one specific (ξ-th) trajectory of the AES:

$$\{^{\xi}a(t)\}, \quad 0 < t < {}^{\xi}T, \tag{15}$$

where ξ is the number of the possible direction of the wind velocity vector, which takes all values of some interval of the numerical axis from 1 to s, where s is the number of possible wind directions.

Thus, for all trajectories of the AES

$$\{^{\xi}A(t)\}, \quad \{^{\xi}h(t)\}, \quad 0 < t < {}^{\xi}T$$

the random process of change of the aerodynamic coefficient can be represented by time functions $\{^{k}a(t)\}$, $0 < t < {}^{\xi}T$ and can be described by its statistical characteristics.

The most universal statistical characteristics for random processes are their correlation functions. For random processes $\{^{k}a(t)\}$ the correlation function can be written in the form

$$R_{aa}(t_1, t_2) = \langle {}^{k}a(t_1)\,{}^{k}a(t_2)\rangle \text{ av. index} = \lim_{N \to \infty} \frac{\sum_{k=1}^{N} {}^{k}a(t_1)\,{}^{k}a(t_2)}{N}. \tag{16}$$

For each realization $^{k}a(t)$ for a fixed k we take the values of $^{k}a(t_1)$ and $^{k}a(t_2)$ and multiply. After this all such products obtained for different realizations are added and divided by the number of realizations. The result thus depends on the selection of times t_1 and t_2.

We break up set $\{^{k}a(t)\}$ into subsets $\{a(t)\}$, $0 < t < {}^{\xi}T$ such that each of the realizations entering a given subset has the same period of reception $^{\xi}T$. The number of such subsets is equal to the number of trajectories u considered.

We examine on realization (recording) $^{\xi}a(t)$ of finite length $^{\xi}T$ of random process $\{^{\xi}a(t)\}$. We divide this recording into n identical intervals by points equidistant from one another with a distance of Δt between them and denote the values obtained of the function at these points a_1, $a_2, \ldots, a_n$. Then the following formula will give the approximate value of the correlation function $^{\xi}R_{aa}(\tau)$ (in this case the origin of the interval of realization $^{\xi}T$ is taken as the origin of the coordinates):

$$^{\xi}R_{aa}(\tau) \approx {}^{\xi}R_{aa}(\Delta tm) = \frac{1}{{}^{\xi}T - \Delta tm} \int_{0}^{{}^{\xi}T - \Delta tm} {}^{\xi}a(t)\,{}^{\xi}a(t + \Delta tm)\,dt \approx \frac{1}{{}^{\xi}T - \Delta tm} \sum_{i=1}^{i=n-m} {}^{\xi}a(t_i)\,{}^{\xi}a(t_i + \Delta tm)\,\Delta t, \tag{17}$$

$$\Delta t = \frac{^{\xi}T}{n}, \quad \tau = \Delta tm, \quad m = 0, 1, 2, \ldots, n.$$

Then

$$^{\xi}R_{aa}(m) \approx \frac{1}{n - m} \sum_{i=1}^{i=n-m} {}^{\xi}a_i\,{}^{\xi}a_{i+m}, \tag{18}$$

where

$$^{\xi}a_i = {}^{\xi}a(t_i), \quad {}^{\xi}a_{i+m} = {}^{\xi}a(t_i + \Delta tm).$$

To obtain $^\xi R_{aa}$ (m) we must calculate n − m products $a_1 a_{1+m}$, $a_2 a_{2+m}$, . . . by formula (18), shifting each value by Δtm units and multiplying the original values by the corresponding values obtained after the shift. Then the n − m products are added and their sum is divided by the number of terms.

We write several of their sums:

$$n\,^\xi R_{aa}(0) = a_1^2 + a_2^2 + \ldots + a_{n-1}^2 + a_n^2,$$
$$(n-1)\,^\xi R_{aa}(1) = a_1 a_2 + a_2 a_3 + \ldots + a_{n-2}a_{n-1} + a_{n-1}a_n,$$
$$(n-1)\,^\xi R_{aa}(-1) = a_1 a_2 + a_2 a_3 + \ldots + a_{n-2}a_{n-1} + a_{n-1}a_n,$$
$$(n-2)\,^\xi R_{aa}(2) = a_1 a_3 + a_2 a_4 + \ldots + a_{n-3}a_{n-2} + a_{n-2}a_n, \qquad (19)$$
$$(n-2)\,^\xi R_{aa}(-2) = a_1 a_3 + a_2 a_4 \ldots + a_{n-\varsigma}a_{n-2} + a_{n-2}a_n,$$
$$\cdot \cdot \cdot \cdot \cdot \cdot \cdot \cdot \cdot \cdot \cdot \cdot \cdot \cdot \cdot \cdot \cdot \cdot \cdot \cdot \cdot \cdot \cdot \cdot \cdot \cdot$$
$$(n-m)\,^\xi R_{aa}(m) = a_1 a_{m+1} + a_2 a_{m+2} + \ldots + a_{n-1}a_{n-(m+1)} + a_{n-m}a_n.$$

From an examination of system of equations (19) we can conclude that $^\xi R_{aa}(\tau)$ has the following two properties:

$$1)\ ^\xi R_{aa}(\tau) = \,^\xi R_{aa}(-\tau)\ \text{for all}\ \ m = 1, 2, 3, \ldots, n, \qquad (20)$$

i.e., the correlation function of the change of the aerodynamic coefficient is an even function;

$$2) \qquad \sum_{i=1}^{n-m} (a_i \pm a_{i+m})^2 > 0,$$

but

$$\sum_{i=1}^{n-m} (a_i \pm a_{i+m})^2 = \sum_{i=1}^{n-m} a_i^2 \pm 2 \sum_{i=1}^{n-m} a_i a_{i+m} + \sum_{i=1}^{n-m} a_{i+m}^2 = n\,^\xi R_{aa}(0) \pm (n-m)\,^\xi R_{aa}(m) + n\,^\xi R_{aa}(0).$$

Consequently,

$$^\xi R_{aa}(0) > \left| \frac{n-m}{n}\,^\xi R_{aa}(m) \right|.$$

When $\Delta t \to 0$ and n → ∞ for all finite parameters of the correlation

$$\lim_{n \to \infty} \frac{n-m}{n} = 1.$$

Therefore,

$$^\xi R_{aa}(0) > |\,^\xi R_{aa}(\tau)|. \qquad (21)$$

It follows from expression (21) that the correlation function is symmetric relative to point $\tau = 0$. If now we assume that the number of realizations $^\xi a$ (t) is equal to s and the duration of each is $^\xi T$, the correlation function can be represented by the formula

$$^\xi R_{aa}(m) = \frac{1}{s(n-m)} \sum_{\substack{i=1 \\ \xi=1}}^{\substack{\xi=s \\ i=n-m}} {}^\xi a_i\,{}^\xi a_{i+m}. \qquad (22)$$

Further, on averaging the set of recordings of random process $\{^k a(t)\}$, $0 < t < {}^\xi T$, we obtain the general formula for calculating the correlation function of the change of the aerodynamic-moment coefficient for a certain group of trajectories of the AES with an equiprobable azimuthal wind direction at the start of communication (time of appearance of the object from behind the horizon t = 0, h = 0);

$$R_{aa}(\tau) = \frac{1}{u} \sum_{\xi=1}^{\xi=u} {}^\xi R_{aa}(\tau), \tag{23}$$

where u is the number of trajectories of the AES considered.

4. Statistical Characteristics of the Moment on the Control Axle of the Antenna Guidance System

Knowing the correlation functions of the random process of change of the aerodynamic-moment coefficient and of the process of fluctuations of wind velocity, we determine the correlation function of the moment acting on the control axle of the antenna. On the basis of expression (6):

$$^k M(t) = {}^k a(t) [\textit{Б} + B\,{}^k v_1(t)], \tag{24}$$

where $\textit{Б} = \langle v^2 \rangle$ and $B = 2v_0$.

Assuming $\{^k a(t)\}$ and $\{^k v_1(t)\}$ are statistically independent, we find the correlation function $R_{mm}(\tau)$ of the total random process of disturbances $\{^k M(t)\}$ applied to the control axle of the guidance system of the radio telescope:

$$R_{mm}(\tau) = \langle {}^k M(t)\,{}^k M(t+\tau)\rangle = \langle \{^k a(t)\,[\textit{Б} + B\,{}^k v_1(t)]\}\{^k a(t+\tau)\,[\textit{Б} + B\,{}^k v_1(t+\tau)]\}\rangle =$$
$$= \langle \{^k a(t)\,{}^k a(t+\tau)\,[\textit{Б}^2 + \textit{Б}B\,{}^k v_1(t) + \textit{Б}B\,{}^k v_1(t+\tau) + B^2\,{}^k v_1(t+\tau)]\}\rangle. \tag{25}$$

Processes $\{^k a(t)\}$ and $\{^k v_1(t)\}$ are not cross-correlated, and therefore their average over the set can be examined separately:

$$R_{mm}(\tau) = \langle {}^k a(t)\,{}^k a(t+\tau)\rangle \langle \textit{Б}^2 + \textit{Б}B\,{}^k v_1(t) + \textit{Б}B\,{}^k v_1(t+\tau) + B^2\,{}^k v_1(t)\,{}^k v_1(t+\tau)\rangle,$$
$$\langle \textit{Б}B\,{}^k v_1(t)\rangle = \langle \textit{Б}B\,{}^k v_1(t+\tau)\rangle = 0.$$

Since the mathematical expectation of the fluctuations of wind velocity relative to the average value of wind velocity equals zero,

$$R_{mm}(\tau) = \langle {}^k a(t)\,{}^k a(t+\tau)\rangle \langle \textit{Б}^2 + B^2\,{}^k v_1(t)\,{}^k v_1(t+\tau)\rangle,$$
$$\langle {}^k a(t)\,{}^k a(t+\tau)\rangle = R_{aa}(\tau),$$
$$\langle B^2\,{}^k v_1(t)\,{}^k v_1(t+\tau)\rangle = R_{v_1 v_1}(\tau)\,B^2.$$

Then

$$R_{mm}(\tau) = R_{aa}(\tau)\,[\textit{Б}^2 + B^2 R_{v_1 v_1}(\tau)]. \tag{26}$$

5. Example of Calculation

We will examine an example of the calculation of the statistical characteristics of random loads on the guidance system of the radio telescope with respect to the azimuth for three possible cases of movement of the antenna:

1) for tracking the trajectory of an AES, $a_1 = a(t)$;

2) for rotating at a constant maximum speed with respect to the azimuth $\omega_A(t) = \Omega_A$ at an angle of elevation equal to zero, $a_2 = a(t)$; and,

3) for the position of the antenna in space remaining constant, with

$$a_3 = a_{max} = \text{const.}$$

We denote the correlation functions of the aerodynamic coefficient on the control axle for each of the cases analyzed, respectively

$$R_{1aa}(\tau), \quad R_{1mm}(\tau),$$
$$R_{2aa}(\tau), \quad R_{2mm}(\tau),$$
$$R_{3aa}(\tau), \quad R_{3mm}(\tau).$$

The method examined above permits determining the correlation functions $R_{1aa}(\tau)$ and $R_{2aa}(\tau)$.

Determination of $R_{1aa}(\tau)$.   Data for the calculation: circular orbit of the AES; $H = 200$ km; $\Omega_A = 3$ deg/sec is the maximum azimuthal tracking speed at the point of culmination (see Figs. 7 and 8); $R_E = 6.37 \cdot 10^6$ m is the earth's radius; $q = v^2/(R_E + H)$ is the gravitational acceleration; $v = g\sqrt{g(R_E + H)} = 8.04 \cdot 10^3$ m/sec is the linear velocity of rotation of the AES around the earth; $\omega_\alpha = v/R_E = 0.07$ deg/sec is the angular velocity of rotation of the AES around the earth; $\cos\varphi = \omega_\alpha/\Omega_A = 0.023$ is the cosine of the angle of inclination of the orbital plane to the horizontal plane at the observation point

$$\alpha_0 = \sin^{-1}\left[\frac{R_E}{R_E + H}\frac{1}{\sqrt{1 - \left(\frac{\omega_\alpha}{\Omega_A}\right)^2}}\right] = 76°;$$

and

$$T = \frac{180° - 2\alpha}{\omega_\alpha} = 400$$

is the period of reception

$$A(t) = \tan^{-1}[0.0233\tan(0.07t + 76°)], \tag{27}$$

$$h(t) = \tan^{-1}\frac{\sin(0.07t + 76°) - 0.97}{\sqrt{\cos^2(0.07t + 76°) + 0.54\cdot10^{-3}\sin(0.07t + 76°)}};$$

$$\omega_A = \frac{dA(t)}{dt} = \frac{1.63\cdot10^{-3}}{\cos^2(0.07t + 76°) + 0.54\cdot10^{-3}\sin(0.07t + 76°)},$$

$$\omega_h = \frac{dh(t)}{dt} = 1.73\sqrt{\omega_A}\frac{\cos(0.07t + 76°) + 307\,\omega_A\sin(28° - 0.14t)[\sin(0.07t + 76°) - 0.97]}{1 + 613\,\omega_A[\sin(0.07t + 76°) - 0.97]^2}. \tag{28}$$

The graphs of $A(t)$, $h(t)$, $\omega_A(t)$, and $\omega_h(t)$ on the basis of formulas (27) and (28) are shown in Figs. 7 and 8. The correlation function $R_{1aa}(\tau)$ was calculated according to (22):

$$R_{1aa}(m) \approx \frac{1}{8(40 - m)}\sum_{\substack{i=0 \\ \xi=1}}^{\substack{\xi=8 \\ i=40-m}}{}^\xi a_i{}^\xi a_{i+m},$$

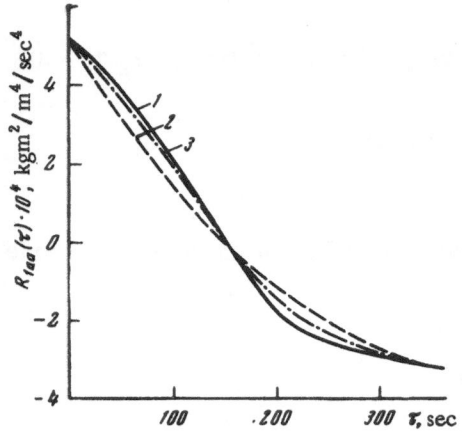

Fig. 4. Correlation function of aerodynamic-moment coefficient and its analytic approximations for tracking of an AES. 1) Experimental curve; 2) approximation according to (29); 3) approximation according to (30).

where $\tau = m\Delta t$ is a parameter of the correlation; $T = n\Delta t = 400$ sec; $\Delta t = 10$ sec; $n = 40$; and $s = 8$ is the number of possible wind directions at the start of tracking (during tracking of the antenna it was assumed that the wind direction remains constant).

In conformity with the curves of the wind-tunnel test (see Fig. 1)

when $\xi = 1$	$^1A = 0$,
when $\xi = 2$	$^2A = 45°$,
when $\xi = 3$	$^3A = 90°$,
$\cdots$	$\cdots\cdots$
when $\xi = 8$	$^8A = 315°$.

The graph of the correlation function $R_{1aa}(\tau)$ is presented in Fig. 4. Two analytic approximations of the function are given in Fig. 4:

$$R_{1aa}(\tau) = D\,e^{-\mu(\tau)}\cos\beta\tau - R_0, \qquad (29)$$

$$R_{1aa}(\tau) = D\,e^{-\mu(\tau)}(\cos\beta\tau + k\sin\beta_1|\tau|) - R_0), \qquad (30)$$

where β is the resonance frequency; μ is a parameter of decay; and R_0 is the square of the mean value of the aerodynamic-moment coefficient during tracking. Expression (30) gives a better approximation to the calculated curve 1 (see Fig. 4, curve 3) than (29), and therefore we take expression (30) for the analytic approximation of $R_{1aa}(\tau)$. After substitution of the numerical values

$$R_{1aa}(\tau) = -32.2\cdot10^3 + 83.8\cdot10^3\,e^{-0.048|\tau|}(\cos0.044\,\tau + 0.1\sin0.02\,|\tau|). \qquad (31)$$

Determination of $R_{2aa}(\tau)$. Data for calculation: $\omega_A(t) = \Omega_A = 3$ deg/sec, $h(t) = 0°$.

The correlation function $R_{2aa}(\tau)$ was calculated according to (22):

$$R_{2aa}(m) = \frac{1}{n-m}\sum_{i=0}^{i=n-m} a_i a_{i+m},$$

where $\tau = m\Delta t$, $\Delta t = 1$ sec, $n = 120$ and corresponds to a complete rotation of the antenna azimuthally by 360°.

The calculation of $R_{2aa}(\tau)$ was performed in the interval T of one recording (realization) $a(t)$ obtained after change of variable A by t in the wind-tunnel graph (see Fig. 2):

$$\frac{A}{\Omega_A} = t, \quad T = \frac{360°}{3\text{ deg/sec}} = 120\text{ sec}.$$

The graph of $R_{2aa}(\tau)$ is shown in Fig. 5 (curve 1). Its two analytic approximations are given in the same figure:

$$R_{2aa}(\tau) = D\,e^{-\mu|\tau|}\cos\beta\tau, \qquad (32)$$

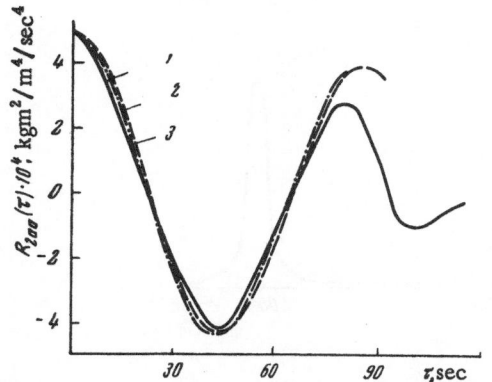

Fig. 5. Correlation function of the aero-dynamic coefficient and its analytic approximations for movement of the antenna with $\omega_A = 3$ deg/sec, $n = 0°$. 1) Experimental curve; 2) approximation according to (33); 3) approximation according to (32).

$$R_{2aa}(\tau) = D\, e^{-\mu|\tau|}\left[\cos\beta\tau + \frac{\mu}{\beta}\sin\beta\,|\tau|\right]. \quad (33)$$

Comparing the curves of the analytic approximation of the correlation function $R_{2\,aa}(\tau)$ with the experimental curve, we see that the first wave of the experimental curve is nicely approximated by a decaying exponential cosine function (32). At the end of the experimental curve we observe random fluctuations around zero, which can be explained by the selection of a finite length of the recording in place of an infinite length and by the weak correlation for large parameters of correlation τ:

$$R_{2aa}(\tau) = 4.95 \cdot 10^4\, e^{-0.00314\,|\tau|} \cos 0.073\,\tau.$$

Correlation Function $R_{3\,aa}(\tau) = a_{max}^2$.

We set $a_{max} = 400$ kgm/$(\text{m/sec})^2$. The correlation function of the wind velocity fluctuations has the form of (7). The range of frequencies of the change of the spectral density of the fluctuations is from 0.4 to 3 sec^{-1} [1,3]. For our problem we take $\nu = 1\ \text{sec}^{-1}$. For the average wind speed $v_0 = 15$ m/sec (according to [1]) the standard deviation from the average wind speed is $0.5v_0$, then

$$[\langle v_1^2\rangle]^{1/2} = 0.5 \cdot 15 = 7.5\,\text{m/sec}.$$

Accordingly the mean square of the wind-speed fluctuations is

$$\langle v_1^2(t)\rangle = 56.25\,(\text{m/sec})^2,$$

and

$$R_{v_1 v_1}(\tau) = 56.25\, e^{-|\tau|}. \quad (34)$$

On the basis of (26), (30), (33), and (34) we determine the correlation functions of the moments:

$$R_{1mm}(\tau) = -M_1^2 + D_1 e^{-\mu_1|\tau|}\cos\beta_1\tau + D_2 e^{-\mu_1|\tau|}\sin\beta_2\,|\tau| + D_3 e^{-\mu_2|\tau|}\cos\beta_1\tau + D_4 e^{-\mu_2|\tau|}\sin\beta_2\,|\tau| - D_5 e^{-\nu|\tau|},$$

$$R_{2mm}(\tau) = D_6 e^{-\mu_3|\tau|}\cos\beta_3\tau + D_7 e^{-\mu_4|\tau|}\cos\beta_3\tau,$$

$$R_{3mm}(\tau) = M_2^2 + D_8 e^{-\nu|\tau|},$$

where $M_1^2 = 25.4 \cdot 10^8\,(\text{kgm})^2$, $\mu_1 = 0.0048\,\text{sec}^{-1}$

and

$$
\begin{array}{ll}
M_2^2 = 126 \cdot 10^8\ (\text{kgm})^2, & \mu_2 = 1.0048\ \text{sec}^{-1}, \\
D_1 = 66.2 \cdot 10^8\ (\text{kgm})^2, & \nu = 1\ \text{sec}^{-1}, \\
D_2 = 6.62 \cdot 10^8\ (\text{kgm})^2, & \mu_3 = 0.00314\ \text{sec}^{-1}, \\
D_3 = 42.4 \cdot 10^8\ (\text{kgm})^2, & \mu_4 = 1.00314\ \text{sec}^{-1}, \\
D_4 = 4.24 \cdot 10^8\ (\text{kgm})^2, & \beta_1 = 0.00437\ \text{sec}^{-1}, \\
D_5 = 16.3 \cdot 10^8\ (\text{kgm})^2, & \beta_2 = 0.021\ \text{sec}^{-1}, \\
D_6 = 39.1 \cdot 10^8\ (\text{kgm})^2, & \beta_3 = 0.073\ \text{sec}^{-1}. \\
D_7 = 25.1 \cdot 10^8\ (\text{kgm})^2, &
\end{array}
$$

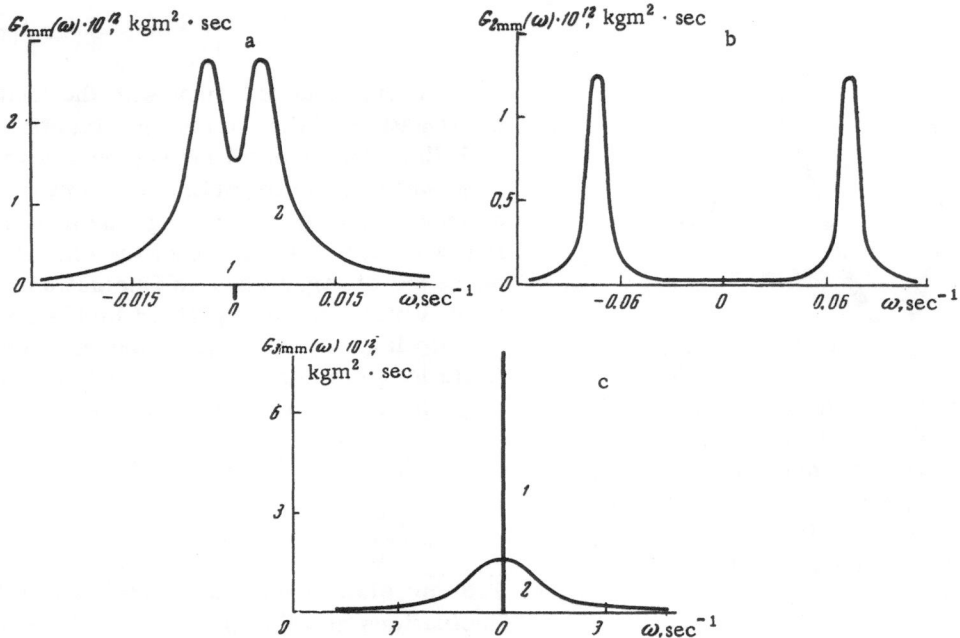

Fig. 6. Spectral densities of the moment on the azimuthal control axle. a) When tracking an AES (1 is the constant component, 2 is the variable component); b) during movement with $\omega_A = 3$ deg/sec, n = 0°; c) for $\omega_A = 0$, $\omega_n = 0$, and $\omega = a_{max}$ (1 is the constant component, 2 is the variable component).

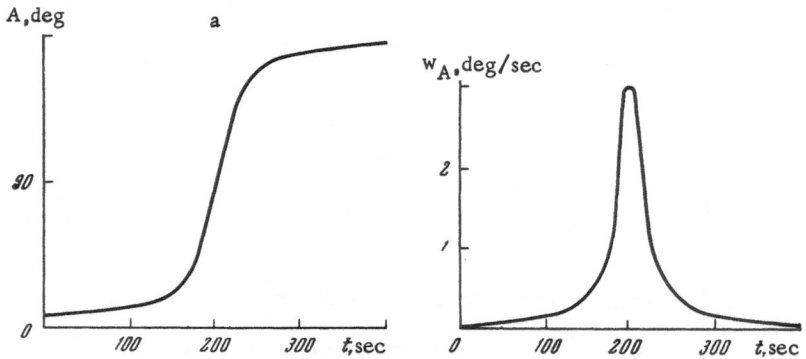

Fig. 7. Time dependence of angle of azimuth (a) and of azimuthal speed (b).

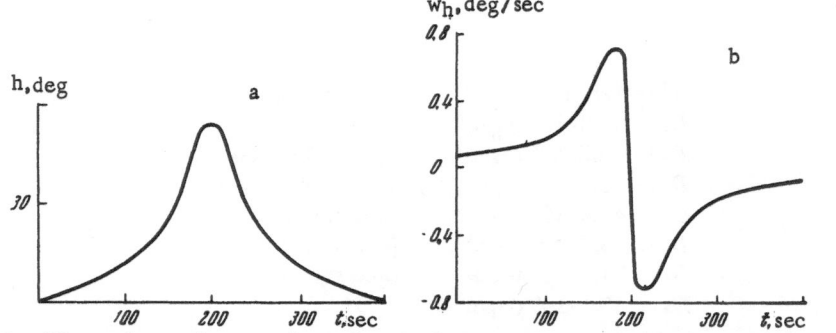

Fig. 8. Time dependence of angle of elevation (a) and elevation speed (b).

From the known correlation functions we determine the spectral density of the moment on the control axle. For an arbitrary random process $\{^k a(t)\}$ when $T \to \infty$ we have [2]

$$G_{aa}(f) = \lim_{T \to \infty} \int_{-\infty}^{\infty} e^{-j2\pi f \tau} \left[\frac{1}{2T} \int_{-T}^{T} R_{aa}(t_1 t + \tau) \, dt \right] d\tau. \tag{35}$$

For correlation function $R_{1aa}(\tau)$ when the antenna is tracking an AES moving over a stationary circular orbit, the operation $\frac{1}{2T} \int_{-T}^{T} R_{aa}(t_1 t + \tau) \, dt$ was performed according to formula (23).

Correlation functions $R_{2aa}(\tau)$ and $R_{3aa}(\tau)$ correspond to stationary random processes. Consequently, formula (35) can be rewritten in the form

$$G_{aa}(f) = 2 \int_{-\infty}^{\infty} R_{aa}(\tau) e^{-j2\pi f \tau} \, d\tau. \tag{36}$$

If we use the angular frequency ω [sec^{-1}] in place of f [Hz], expression (36) takes the form of the known Wiener — Khinchine relation

$$G_{aa}(\omega) = \frac{2}{\pi} \int_{0}^{\infty} R_{aa}(\tau) \cos \omega \tau \, d\tau. \tag{37}$$

We note that $G(f)$ is a double Fourier transform of $R_{aa}(\tau)$.

Thus, the general expression of the spectral density of the fluctuations of the moment on the control axle from the wind load is

$$G_{mm}(\omega) = \frac{2}{\pi} \int_{0}^{\infty} R_{mm}(\tau) \cos \omega \tau \, d\tau. \tag{38}$$

Then the spectral density of the moments corresponding to the correlation functions $R_{1mm}(\tau)$, $R_{2mm}(\tau)$, and $R_{3mm}(\tau)$ have the form

$$G_{1mm}(\omega) = -2\pi_1 M_1^2 \delta(\omega) + \mu_1 D_1 \left[\frac{1}{\mu_1^2 + (\beta_1 - \omega)^2} + \frac{1}{\mu_1^2 + (\beta_1 + \omega)^2} \right] + D_2 \left[\frac{\beta_2 - \omega}{\mu_1^2 + (\beta_2 - \omega)^2} + \frac{\beta_2 + \omega}{\mu_1^2 + (\beta_2 + \omega)^2} \right]$$

$$+ \mu_2 D_3 \left[\frac{1}{\mu_2^2 + (\beta_1 - \omega)^2} + \frac{1}{\mu_2^2 + (\beta_1 + \omega)^2} \right] + D_4 \left[\frac{\beta_2 - \omega}{\mu_2^2 + (\beta_2 - \omega)^2} + \frac{\beta_2 + \omega}{\mu_2^2 + (\beta_2 + \omega)^2} \right] - \frac{2 D_5 v}{v^2 + \omega^2},$$

$$G_{2mm}(\omega) = \mu_3 D_6 \left[\frac{1}{\mu_3^2 + (\beta_3 - \omega)^2} + \frac{1}{\mu_3^2 + (\beta_3 + \omega)^2} \right] + \mu_4 D_7 \left[\frac{1}{\mu_4^2 + (\beta_3 - \omega)^2} + \frac{1}{\mu_4^2 + (\beta_3 + \omega)^2} \right], \tag{39}$$

$$G_{3mm}(\omega) = 2\pi M_2^2 \delta(\omega) + \frac{2\pi D_8 v}{v^2 + \omega^2},$$

where $\delta(\omega)$ is the unit impulse function.

After substituting the numerical values we obtain

$$G_{mm}(\omega) = -159 \cdot 10^8 \delta(\omega) + 0.318 \cdot 10^8 \left[\frac{1}{0.23 \cdot 10^{-4} + (0.00437 - \omega)^2} + \right.$$

$$\left. + \frac{1}{0.2 \cdot 10^{-4} + (0.00437 + \omega)^2} \right] + 6.62 \cdot 10^8 \left[\frac{0.021 - \omega}{0.23 \cdot 10^{-4} + (0.021 - \omega)^2} + \right.$$

$$\left. + \frac{0.021 - \omega}{0.23 \cdot 10^{-4} + (0.021 + \omega)^2} \right] + 42.4 \cdot 10^8 \left[\frac{1}{1 + (0.00437 - \omega)^2} + \right.$$

$$+ \frac{1}{1+(0.00437+\omega)^2}\Big] + 4.24 \cdot 10^8 \Big[\frac{0.021-\omega}{1+(0.021-\omega)^2} + \frac{0.021-\omega}{1+(0.021+\omega)^2}\Big] - \frac{32.6}{1+\omega^2} ,$$

$$G_{2\,mm}(\omega) = 0.123 \cdot 10^8 \Big[\frac{1}{9.85 \cdot 10^{-6}+(0.073-\omega)^2} + \frac{1}{9.85 \cdot 10^{-6}+(0.073+\omega)^2}\Big] +$$

$$+ 25.1 \cdot 10^8 \Big[\frac{1}{1+(0.073-\omega)^2} + \frac{1}{1+(0.073+\omega)^2}\Big] ,$$

$$G_{3\,mm}(\omega) = 7.91\,\delta(\omega) + \frac{161}{1+\omega^2} .$$

The curves of the change of spectral densities $G_{1\,mm}(\omega)$, $G_{2\,mm}(\omega)$, and $G_{3\,mm}(\omega)$ according to (39) are presented in Fig. 6a, b, c.

Integration of the spectral density of the moment with respect to all frequencies gives the mean square of the moment in $(kgm)^2$ (Fig. 7):

$$\langle M_1^2(t)\rangle = M_1^2 + D_1 + D_2 + D_3 + D_4 - D_5 = 66.9 \cdot 10^8,$$
$$\langle M_2^2(t)\rangle = D_6 + D_7 = 64.2 \cdot 10^8,$$
$$\langle M_3^2(t)\rangle = M_2^2 + D_8 = 206.6 \cdot 10^8.$$

Accordingly the mean square values of the moments from the wind load on the control axle of the azimuthal drive are equal to: when tracking an object $[\langle M_1^2(t)\rangle]^{\frac{1}{2}} = 82 \cdot 10^3$ kgm; during movement with a constant maximum azimuthal speed $[\langle M_2^2(t)\rangle]^{\frac{1}{2}} = 80 \cdot 10^3$ kgm; with a stationary antenna $[\langle M_3^2(t)\rangle]^{\frac{1}{2}} = 142 \cdot 10^3$ kgm (Fig. 8).

Conclusions

1. Random disturbances acting on the guidance system of a radio telescope are determined both by fluctuations of the wind and by the random character of change of the aerodynamic coefficient of the antenna.

2. A method is given for determining the statistical characteristics of change of the aerodynamic-moment coefficient based on the results of wind-tunnel tests on a model radio telescope.

3. The correlation functions of the moments on the control axle of the guidance system drive are decaying exponential-cosine functions.

4. When solving problems of determining the statistical characteristics of the disturbances, we disregarded the filtering properties of the antenna with respect to wind speed fluctuations. It was considered that the energy of the wind is transformed linearly by the antenna without losses. Therefore, the results obtained in the work are overstated with respect to absolute values, which obviously can be a useful margin when designing.

Consideration of the antenna as an aerodynamic filter of wind varying in magnitude is a complex independent problem.

APPENDIX

Kinematics of Tracking an Artificial Earth

Satellite by Radio Telescope

We will examine the kinematics of tracking an AES by radio telescope on the basis of the following assumptions: the earth is a sphere; the orbit of the satellite is circular; and the orbital plane passes through the center of the earth.

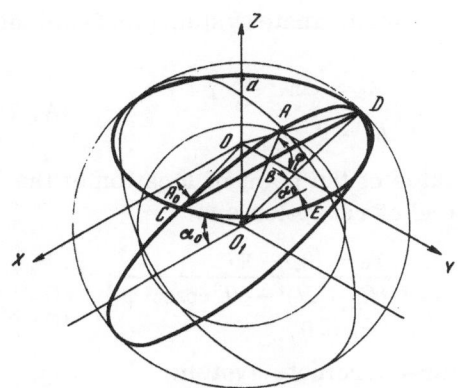

Fig. 9. Diagram for determining the laws of motion of the antenna when tracking an AES.

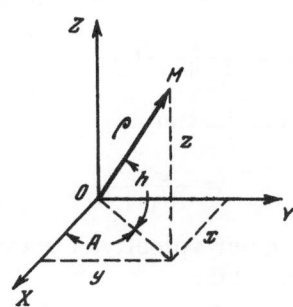

Fig. 10. Diagram of the spherical-coordinate system of the radio telescope.

We call R_E the earth's radius; H is the altitude of the satellite above the earth's surface; v is the linear velocity of rotation of the satellite around the earth (Fig. 9).

The spherical-coordinate system of the radio telescope (Fig. 10) is

$$x = \rho \cos h \cos A,$$
$$y = \rho \cos h \sin A,$$
$$z = \rho \sin h,$$
$$\rho = \sqrt{x^2 + y^2 + z^2},$$
$$A = \tan^{-1} \frac{y}{x},$$
$$h = \tan^{-1} \frac{z}{\sqrt{x^2 + y^2}}.$$

We find the equation of the trajectory of motion of the satellite relative to coordinate system X, Y, Z, the origin of which coincides with the observation point.

Equation of a sphere of radius R_E + H:

$$x^2 + y^2 + (z - R_E)^2 = (R_E + H)^2. \qquad (A.1)$$

Equation of plane CAD:

$$z = y \tan \varphi - R_E. \qquad (A.2)$$

Equation of a sphere:

$$x^2 + y^2 + (z - R_E)^2 = R_E^2. \qquad (A.3)$$

Equation of the horizontal plane:

$$z = 0. \qquad (A.4)$$

Equation of the satellite orbit:

$$x^2 + y^2 + (z - R_E)^2 = (R_E + H)^2,$$
$$z = y \tan \varphi - R_E. \qquad (A.5)$$

Points of intersection of the horizontal plane by the satellite orbit

$$x_0^2 + y_0^2 + (z_0 - R_E)^2 = (R_E + H)^2,$$
$$z_0 = y_0 \tan \varphi - R_E,$$
$$z_0 = 0.$$

Hence we get

$$y_0 = R_E \cot \varphi,$$
$$x_0^2 = 2R_E H + H^2 - R_E^2 \cot^2 \varphi,$$
$$x_0 = \sqrt{(R_E + H)^2 - R_E^2 \operatorname{cosec}^2 \varphi},$$
$$z_0 = 0. \qquad (A.6)$$

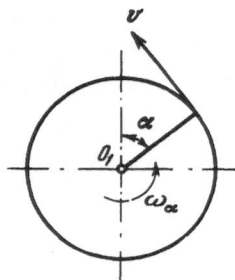

Fig. 11. Diagram for the equation of motion of the satellite.

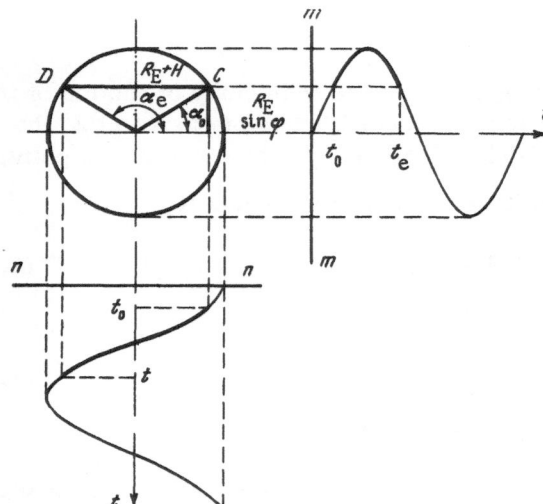

Fig. 12. Diagram for the determination of the projections of the motion of the satellite onto two mutually perpendicular directions.

In a polar-coordinate system this becomes

$$A_0 = \tan^{-1} \frac{y_0}{x_0} ,$$
$$h_0 = 0. \tag{A. 7}$$

The coordinates of the point of location of the AES at the end of reception are

$$y_e = R_E \cot \varphi,$$
$$x_e = - \sqrt{(R_E + H)^2 - R_E^2 \operatorname{cosec}^2 \varphi},$$
$$z_e = 0 \tag{A. 8}$$

or, in a polar-coordinate system:

$$A_e = \tan^{-1} \frac{y_e}{x_e} = \tan^{-1} \left(- \frac{y_0}{x_0} \right) = \pi - \tan^{-1} \frac{y_0}{x_0} ,$$
$$h_e = 0. \tag{A. 9}$$

Where the equation of motion of the satellite relative to the center of the earth is (Fig. 11)

$$\alpha (t) = \frac{v}{R_E + H} t, \tag{A.10}$$

the period of communication is

$$T = \frac{\alpha_e - \alpha_0}{\omega_\alpha} = \frac{\alpha_e - \alpha_0}{v/(R_E + H)} , \tag{A.11}$$

where ω_α is the angular velocity of rotation of the AES around the earth.

We find the projections of the motion of the satellite on two mutually perpendicular directions m-m and n-n parallel to the orbital plane of the AES (Fig. 12). Direction m-m belongs to the orbital plane of the AES and is parallel to straight line O_1A:

$$m (t) = (R_E + H) \sin (\omega_\alpha t + \alpha_0),$$
$$\sin (\alpha_0) = \frac{R_E}{R_E + H} \frac{1}{\sin \varphi} , \tag{A.12}$$
$$\alpha_0 = \sin^{-1} \frac{R_E}{R_E + H} \cos \varphi.$$

Direction n-n belongs to the orbital plane and is parallel to the X axis:

$$n (t) = (R_E + H) \cos (\omega_\alpha t + \alpha_0),$$

where n(t) is the projection of the instantaneous position of the AES onto the coordinate axis OX in a rectangular-coordinate system as a consequence of the fact that nn‖CD, and CD‖OX. Direction m-m forms with the horizontal plane an angle φ.

$$y (t) = m (t) \cos \varphi,$$
$$z (t) = m (t) \sin \varphi - R_E, \tag{A.13}$$

where $y(t)$ and $z(t)$ are projections of the instantaneous position of the satellite onto the coordinate axes Y and Z, and φ is the angle formed by the horizontal plane and the orbital plane of the AES.

Thus, the equations of motion of the satellite in projections onto the coordinate axes are

$$
\begin{aligned}
x(t) &= (R_E + H)\cos\left[\frac{v}{R_E + H}t + \sin^{-1}\frac{R_E\,\text{cosec}\,\varphi}{R_E + H}\right], \\
y(t) &= (R_E + H)\cos\varphi\sin\left[\frac{v}{R_E + H}t + \sin^{-1}\frac{R_E\,\text{cosec}\,\varphi}{R_E + H}\right], \\
z(t) &= (R_E + H)\sin\varphi\sin\left[\frac{v}{R_E + H}t + \sin^{-1}\frac{R_E\,\text{cosec}\,\varphi}{R_E + H}\right] - R.
\end{aligned}
\tag{A.14}
$$

In a spherical-coordinate system

$$
\left.
\begin{aligned}
A(t) &= \tan^{-1}[\cos\varphi\tan(\omega_\alpha t + \alpha_0)], \\[4pt]
h(t) &= \tan^{-1}\frac{\sin\varphi\sin(\omega_\alpha t + \alpha_0) - \dfrac{R_E}{R_E + H}}{\sqrt{\cos^2(\omega_\alpha t + \alpha_0) + \cos^2\varphi\sin^2(\omega_\alpha t + \alpha_0)}}, \\[4pt]
\omega_A(t) &= \frac{dA(t)}{dt} = \frac{\omega_\alpha\cos\varphi}{\cos^2(\omega_\alpha t + \alpha_0) + \cos^2\varphi\sin^2(\omega_\alpha t + \alpha_0)}, \\[4pt]
\omega_h(t) &= \frac{dh(t)}{dt} = \frac{\omega_\alpha\left\{\sin\varphi\cos(\omega_\alpha t + \alpha_0)[\cos^2(\omega_\alpha t + \alpha_0) + \cos^2\varphi\sin^2(\omega_\alpha t + \alpha_0)] - \right.}{\{\cos^2(\omega_\alpha t + \alpha_0) + \cos^2\varphi\sin^2(\omega_\alpha t + \alpha_0) + [\sin\varphi\sin(\omega_\alpha t + \alpha_0) -} \\
&\quad \frac{\left. -\frac{1}{2}\sin 2(\omega_\alpha t + \alpha_0)[\cos^2\varphi - 1][\sin\varphi\sin(\omega_\alpha t + \alpha_0) - \frac{R_E}{R_E + H}]\right\}}{-\frac{R_E}{R_E + H}]^2\}\sqrt{\cos^2(\omega_\alpha t + \alpha_0) + \cos^2\varphi\sin^2(\omega_\alpha t + \alpha_0)}}.
\end{aligned}
\right\}
\tag{A.15}
$$

The azimuth speed reaches its maximum value Ω_A at time $\frac{1}{2}T = (\pi - 2\alpha_0)/2\omega_\alpha$:

$$
\Omega_A = \frac{\omega_\alpha\cos\varphi}{\cos^2\left(\omega_\alpha\frac{T}{2} + \alpha_0\right) + \cos^2\varphi\sin^2\left(\omega_\alpha\frac{T}{2} + \alpha_0\right)} = \frac{\omega_\alpha\cos\varphi}{\cos^2\varphi} = \frac{\omega_\alpha}{\cos\varphi},
$$

$$
\cos\varphi = \frac{\omega_\alpha}{\Omega_A},
$$

from where

$$
\varphi = \cos^{-1}\frac{\omega_\alpha}{\Omega_A}.
$$

Then expression (A.15) is rewritten as

$$
\omega_A(t) = \frac{\omega_\alpha^2}{\Omega_A}\frac{1}{\cos^2(\omega_\alpha t + \alpha_0) + \dfrac{\omega_\alpha^2}{\Omega_A^2}\sin^2(\omega_\alpha t + \alpha_0)},
$$

$$
\omega_h(t) = \frac{\omega_\alpha\left\{\dfrac{\omega_\alpha^2}{\Omega_A\,\omega_A}\sqrt{1 - \dfrac{\omega_\alpha^2}{\Omega_A^2}}\cos(\omega_\alpha t + \alpha_0) + \dfrac{1}{2}\left[1 - \dfrac{\omega_\alpha^2}{\Omega_A^2}\right]\sin 2(\omega_\alpha t + \alpha_0)\left[\sin(\omega_\alpha t + \alpha_0)\sqrt{1 - \dfrac{\omega_\alpha^2}{\Omega_A^2}} - \dfrac{R_E}{R_E + H}\right]\right\}}{\sqrt{\dfrac{\omega_\alpha^2}{\omega_A\,\Omega_A}}\left\{\dfrac{\omega_\alpha^2}{\omega_A\,\Omega_A} + \left[\sqrt{1 - \dfrac{\omega_\alpha^2}{\Omega_A}}\sin(\omega_\alpha t + \alpha_0) - \dfrac{R_E}{R_E + H}\right]^2\right\}}.
\tag{A.16}
$$

LITERATURE CITED

1. G. C. Newton, L. A. Gould, and J. F. Kaiser, Theory of Linear Servo Systems [Russian translation], Fizmatgiz (1961).
2. J. Bendat, Principles and Applications of Random Noise Theory, John Wiley and Sons, Inc., New York (1958).
3. S. L. Zubkovskii, "Frequency spectrum of fluctuations of the horizontal component of wind velocity in the surface boundary layer," Izv. Akad. Nauk SSSR, Seriya Geofiz., No. 1 (1962).
4. L. G. Loitsyanskii and I. D. Povkh, Aerodynamic Characteristics of Certain Structures [in Russian], Otchet LPI im. Kalinina (1961).
5. V. A. Barbonov and E. A. Rozenman, Vopr. Radioelektron., 5:629 (1963).

A 3-cm NULL RADIOSPECTROMETER FOR OBSERVATIONS
OF THE GALACTIC RADIO LINE OF EXCITED HYDROGEN

V. P. Bibinova, E. V. Borodzich,
R. L. Sorochenko, and I. V. Shavlovskii*

Current radio-astronomical problems require ever greater increases in the sensitivity of radio-astronomical equipment. A great leap in this direction was achieved when low-noise paramagnetic and parametric amplifiers were introduced for radio-astronomical measurement; these amplifiers allowed the radiometer noise level to be reduced drastically. For spectral measurements, however, a simple reduction of the noise level does not always yield a corresponding improvement in the actual sensitivity of the radiometer, since the effects of equipment instabilities and various "parasitic" phenomena increase abruptly. This compels us to seek circuitry which would lower these harmful factors substantially. One of these possibilities was examined in [1], where a null radiospectrometer circuit was advanced. Below, we describe a radiospectrometer operating at a wavelength of 3.3 cm, which was designed on the basis of this circuit and can be used to conduct observations of the spectral radio emission of excited hydrogen. In the spring of 1964, this radiospectrometer [2] detected, for the first time, a cleanly recorded recombination radio line $n_{91} \rightarrow n_{90}$ in the Omega nebula.

1. Basic Operating Principle of the Radiometer

The block diagram of the radiometer is shown in Fig. 1. As described in [1], the basic principle of its operation is based on the joint utilization of two different reception methods: the null method of receiving radio emission in the continuous spectrum, and the differential method of isolating a spectral line against a noise background having a constant spectral density. The radiometer had a broadband amplifier channel ΔF and an additional gain channel having a narrow band Δf which is shifted within the band ΔF when the frequency of the second heterodyne is retuned. When the antenna is attached to a dummy ("equivalent"), a signal with frequency modulation having an amplitude which is proportional to the difference between the antenna and dummy temperatures $T_a - T_d$ (these temperatures are averaged over the band ΔF) is isolated at the output of the broadband channel (after detection by the detector D_1). After appropriate amplification and detection by a synchronous detector, this "error" signal is used to control a compensating-noise signal applied to the dummy, as is the practice in radiometers which operate according to the null method.

On the other hand, the noise from the broadband channel, after detection by the detector D_1, is applied to a balancing cell where it is combined with the noise of the Δf band after it has been detected by the detector D_2.

---

*Deceased.

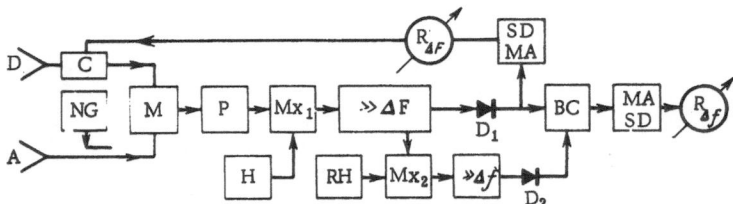

Fig. 1. Block diagram of the null spectral radiospectrometer.
A) Antenna; D) dummy; C) noise compensator; M) modulator (switch);
P) parametric amplifier; ΔF) broadband channel with a 20-MHz band;
Δf) narrowband channel with a 1-MHz band; H) first heterodyne; RH)
retuneable second heterodyne; Mx_1) first mixer; Mx_2) second mixer;
NG) noise generator; BC) balancing cell; MA) modulation–frequency
amplifier; SD) synchronous detector; $R_{\Delta F}$) recording instrument for
the ΔF channel; $R_{\Delta f}$) recording instrument for the Δf channel.

When $\Delta F \gg \Delta f$ and the gain of the Δf channel is such that the values of noise signal applied to the balancing cell from the two bands are equal when the dummy is connected to the receiver input, we will have a signal at the modulation frequency whose amplitude is proportional to the difference $T_a - T_d$ in the Δf band at the output of the cell.

A singly-tuned degenerate parametric amplifier operating in the 3-cm band [3] is used as the rf amplifier of the radiometer. The noise temperature of the radiometer with the parametric amplifier is 425°K with respect to the continuous spectrum. For spectral measurement the noise temperature is doubled and equals 850°K due to the well-known peculiarities of a degenerate parametric amplifier. The bandwidth of the broadband channel is ΔF = 20 MHz, and the bandwidth of the narrowband channel is Δf = 1 MHz. Since the frequency of the first heterodyne was stabilized by means of a high-Q cavity resonator [3], the radiometer provided a measurement and reading accuracy of ±200 kHz with respect to frequency.

In order to record the frequencies of the signals received by the Δf channel when it was tuned over the band, the frequency of the second heterodyne was set at 1 MHz and mixed with an array of reference harmonics generated by a calibrating quartz heterodyne. The frequency markers which were isolated under these conditions were recorded simultaneously with the signal. In order to provide for a constant gain over the tuning range, the Δf channel has a feedback which maintains a constant mean-square noise across the detector D_2.

We will discuss certain specific features of the operation of the null spectral radiometer.

2. The Null Method and the Operation of the
Radiometer over a Continuous Spectrum

We used a ferrite cell, consisting of ferrite operating in a longitudinal field, an absorptive plate, a matching plate, and a magnetization coil, as a dummy antenna with a controllable noise temperature. All of these elements were mounted in a circular waveguide section; one end of the section was coupled with a horn irradiator pointed at the sky, while the other end was attached to a rectangular waveguide which was coupled to the modulator. Since the absorptive plate was oriented parallel to the wide wall of the rectangular waveguide, the radiation from the plate did not enter the radiometer at zero magnetization current, and the temperature of the dummy was produced solely by the received external radiation T_{ex}. When current was present in the coil, the noise temperature of the dummy increased by

$$\Delta T_d = T_0 \sin \varphi + T_{ex} \cos \varphi$$

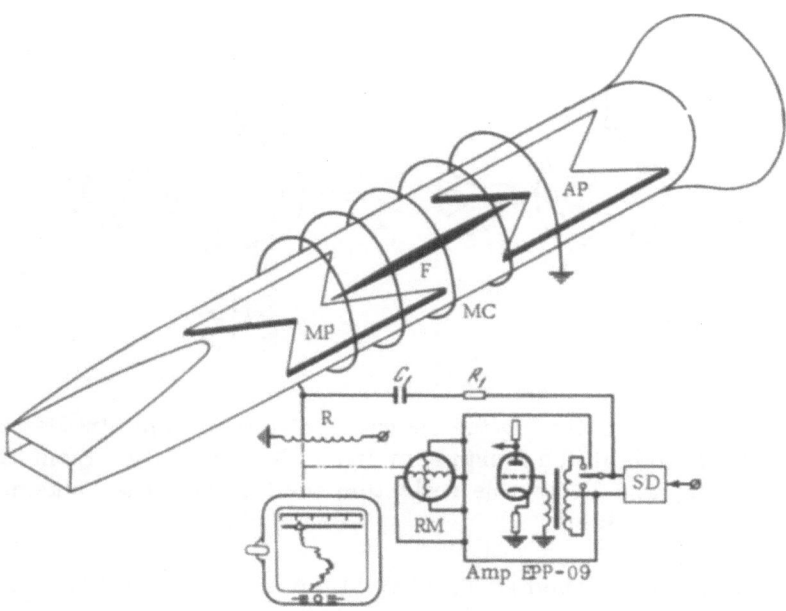

Fig. 2. Block diagram of the servosystem. F) Ferrite; AP) absorptive plate; MP) matching plate; MC) magnetization coil; SD) synchronous detector; R) rheochord; RM) rheochord motor; $R_1 C_1$) first-derivative feedback circuit.

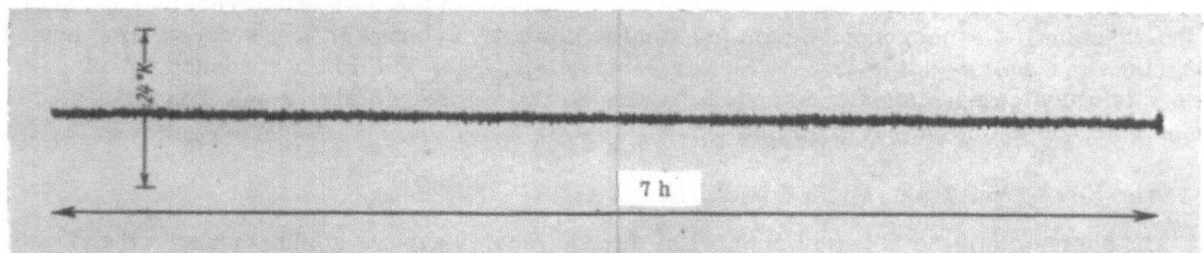

Fig. 3. Example of a recording produced by the null radiometer.

as a result of the rotation of the polarization plane in the ferrite; here T_0 is the temperature of the absorptive rod (the external medium), and φ is the rotation angle of the polarization plane in the ferrite and depends on the magnetization current. For temperature changes of up to 15–20°K, which are usually required in observations, the dependence of ΔT_d on the magnetization current was close to linear. Based on the requirements of the radiospectrometer, the entire dummy system was carefully matched with a standing wave ratio KCB ≤ 1.15 in a 150-MHz band. In order to control the ferrite cell while simultaneously recording the magnetization current, whose magnitude uniquely determined the dummy temperature, we used a standard EPP-09 automatic pen recorder. The error signal isolated across the synchronous detector of the radiometer was applied directly to the input of the EPP amplifier, bypassing the bridge circuit; the control voltage was obtained from a rheochord supplied from an external dc voltage source. For such an arrangement of the feedback circuit the changes of antenna temperature caused the rheochord slider to move until the change in the ferrite magnetization current produced a corresponding increase or decrease in the dummy temperature (the initial control of the system

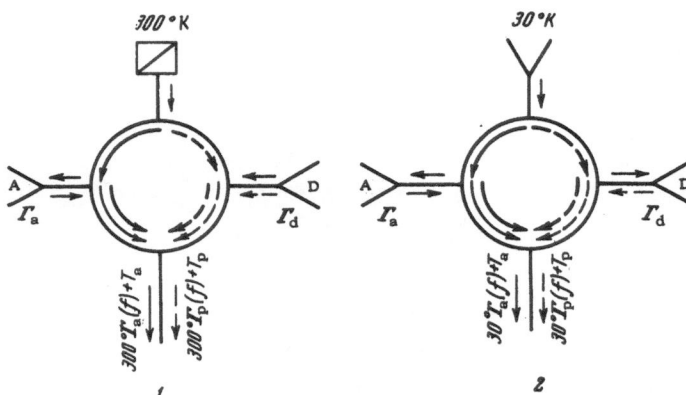

Fig. 4. Diagram illustrating the operation of a four-arm circulator with
(1) a matched mode in the fourth arm and with (2) radiation from the
fourth arm. Γ_a and Γ_d are the reflection coefficients of the antenna and
the dummy, respectively.

for zero current is always established as $T_a \gtrless T_d$ in the absence of a signal from a source of
cosmic radio emissions).

A block diagram of the servosystem is shown in Fig. 2. A control-factor value of Q = 100
was achieved. The oscillatory mode which occurred under these conditions was eliminated by
introducing a first-derivative feedback circuit.

For these parameters the radiometer provided very good fluctuational sensitivity. The
recording of the null line of the radiometer, in the absence of a change in antenna temperature,
was exceedingly stable over a long period of time. Changes in the gain of the receiver channel
and the phase at the synchronous detector output led only to a change in the effective time constant, since in a null radiometer the relationship $\tau_{eff} = \tau/Q$ is valid in accordance with [4],
where τ is the time constant for an open feedback circuit. Figure 3 shows an example of a recording obtained using the null radiometer.

3. Recording of Spectral Signals

The operation of the spectral channel of the radiometer generally substantiated the theoretical calculations fairly well [1]. With a control factor Q = 100 for the radiometer servosystem and additional compensation of the continuous-spectrum noise in the balancing cell to an
accuracy of 10%, the spectral output of the radiometer was sufficiently well protected from intensity changes which were common to the entire band ΔF. Because of this, it was possible to
use a storage time equal to 240^s, and to obtain a fluctuational sensitivity of $\delta T_{\Delta f} = 0.08°K$, with
respect to the spectral channel at a fixed frequency.

However, in tuning this channel over the band the zero line of the output instrument did
not remain constant; it shifted by an amount tens of times greater than the quantity $\delta T_{\Delta f}$ when
the frequency varied within the limits of the analyzed band (10 MHz). Reduction of this "parasitic modulation" effect, which reduced the actual sensitivity substantially, proved to be the
most difficult stage in the process of adjusting the spectrometer. According to the analysis carried out in [1], the parasitic modulation of a null spectral radiometer must be due to three basic
causes:

1) Mismatch of the antenna and dummy channels.

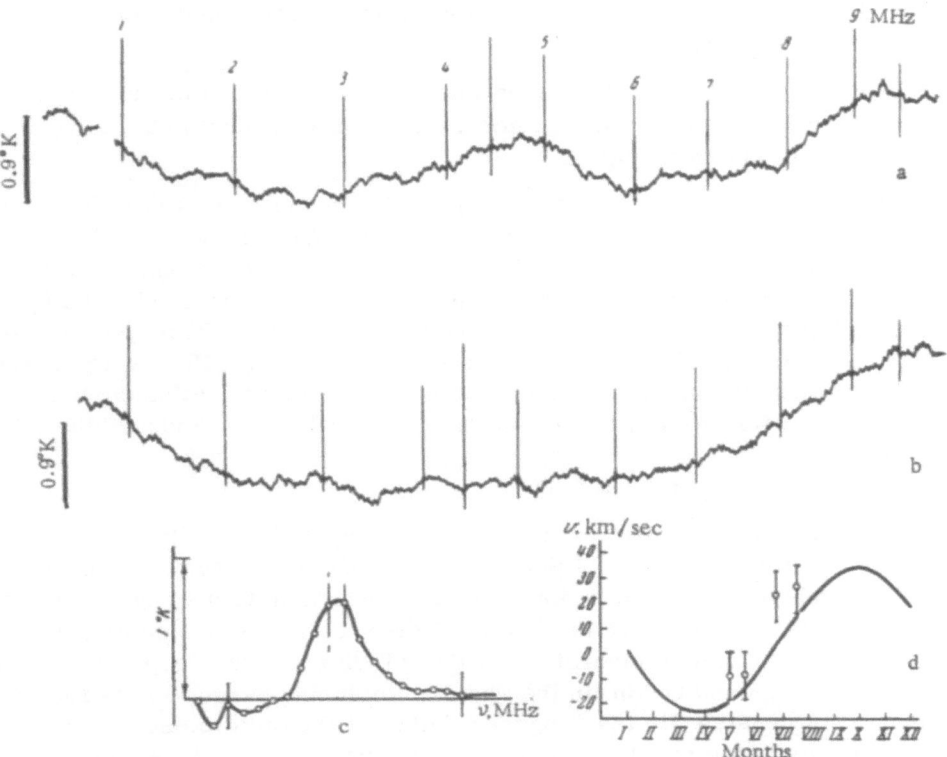

Fig. 5. Results of observing the spectral line of hydrogen in the Omega nebula.
a) Spectrogram of Omega (the short vertical lines are frequency markers
spaced 1 MHz apart, and the long vertical marker is the calculated frequency
of the spectral line); b) control spectrogram for a diverted antenna — the instrumental profile; c) the results of averaging the spectrograms of Omega and five
control spectrograms (the large vertical marker represents the calculated frequency); d) radial velocity of Omega relative to the earth (the solid curve was
calculated, and the points represent experimental data).

2) Nonideal response by the servosystem for operation in the continuous spectrum, combined with a nonuniform gain in the narrowband channel in the ΔF band.

3) Changes in the radiometer's frequency response and in the spectrum of the internal
noise during switching between the antenna and the dummy.

Considering that the servosystem provided fully adequate control, primary attention was
devoted to matching the rf channel and providing reliable decoupling between the switch and the
receiver input. The latter factor is especially essential when a parametric amplifier is used,
since the reaction of the switch on the amplifier affects both the signal frequency and the pumping frequency.

Certain additional causes of parasitic modulation were clarified during the process of
constructing the radiometer. It became obvious from the results of laboratory tests that the
noise radiated by the switch itself makes a substantial contribution to this effect. In our switching system, which used a four-arm circulator in an alternating magnetic field, the 300-degree
noise radiation from the fourth arm entered the antenna and dummy channels. Since the matching of these elements was not ideal, a portion of the noise was reflected and was incident on

the receiver input after having been modulated at the switching frequency due to the difference between the reflection coefficients.

In order to reduce this effect, the matched load in the fourth arm of the circulator was replaced by an auxiliary horn exciter whose temperature corresponded to 30°K. The overall operation of this section is shown in Fig. 4.

As a result of all the measured adopted, the parasitic modulation was reduced substantially, although not eliminated completely. When the radiometer frequency was retuned, the baseline output of the spectral channel was straight but had a certain instrumental profile shaped in the form of a concave curve. A slight straightening of the concavity was achieved by introducing a weak reactance into the auxiliary-horn arm as a compensator. When the slope of the zero line at the output of the spectral channel was reduced to 0.1-0.2 deg/MHz, it was agreed that the parameters obtained for the radio spectrometer were completely satisfactory for the solution of the problem stated; further improvement was not justified by the unavoidable loss of time.

4.   Results of the Observations

The radiometer described above was used in the Spring of 1964 with a 22-m radio telescope of the Physics Institute, Academy of Sciences of the USSR, to carry out observations of the radio line of excited hydrogen in emission nebulae. According to previous calculations [5], we must expect an increase in the spectral density of the received emission by several tenths of a degree relative to the antenna temperature of the RT-22 radio telescope in the direction of the brightest nebulae, Omega and Orion, in the radiometer tuning range centered at 8872.5 MHz. The method of observation consisted in tracking the nebula with the radio telescope and simultaneously tuning the frequency of the spectral channel. For comparison and control purposes the spectrogram for a diverted antenna, which characterized the instrumental profile, was recorded after the spectrogram of the nebula had been recorded. The observations revealed the expected radio line in the Omega nebula. The increase in antenna temperature at the frequency of the center of the line was 0.65°K. Figure 5 shows the spectrograms of the Omega nebula, the control spectrograms, the result of averaging several recordings, and the frequency shift of the observed brightness increase with time of year due to the orbital motion of the earth. The scientific results of the observations have been presented in [2].

Our observations showed the absolute feasibility and necessity of using the null method for the reception of very weak spectral signals. During the process of recording one spectrogram, which took approximately two hours, the level of the received continuous-spectrum noise changed by several degrees due to the insufficient accuracy of source tracking by the telescope, and more greatly (up to 10°K) due to changes in the received radio emission from the atmosphere. Without effective and continuous compensation of the continuous spectrum, the isolation of a spectral signal of tenths of degrees Kelvin would be an unsolvable problem under the stipulated conditions.

LITERATURE CITED

1. E. V. Borodzich and R. L. Sorochenko, Izv. Vuz. Radiofiz., 6:1167 (1963).
2. R. L. Sorochenko and E. V. Borodzich, Dokl. Akad. Nauk SSSR, 163:603 (1965).
3. V. P. Bibinova, A. D. Kuz'min, M. T. Levchenko, V. I. Pushkarev, A. E. Salomonovich, and I. V. Shavlovskii, this volume, p. 129.
4. N. V. Karlov, Candidate's dissertation, P. N. Lebedev Physics Institute, Academy of Sciences of the USSR [in Russian] (1956).
5. R. L. Sorochenko, in: Radio Telescopes (D. V. Skobel'tsyn, ed.), Consultants Bureau, New York (1966), p. 65.

DISCRETE DIODE PHASE SHIFTER
FOR VERY SHORT WAVELENGTHS

S. N. Ivanov and V. T. Solodkov

<u>Introduction</u>

During the last few years semiconductor diodes have been used extensively as switching elements in ultrahigh-frequency technology [1]. They have a number of advantages over switches of the mechanical or electromagnetic types, namely, small size, high reliability, and the capability of providing a good match when switched into a transmission line. These qualities are especially important if a switching system must be devised for a large number of directions.

In the present article a discrete type of diode phase shifter is proposed for the very short wavelength range (30–120 MHz) which introduces a discrete time delay to achieve diffraction control of the antenna beam for a FIAN DKR-1000 north — south radio telescope.

1. Discrete Diode Phase Shifter

The principal requirement of a discrete phase shifter is to provide, with the necessary accuracy, a specified time delay and, at the same time, to match the phase shift closely over the entire frequency range. For a given discrete step of time delay τ the attainable accuracy for the time delay setting is determined by the frequency dispersion of the particular phase shifter stage, which includes a delay line of some length τ_n, by the switching diodes, and by the blocking elements of the control networks.

The type RK-103 cable employed as a delay line in the phase shifter has no frequency dispersion. The frequency dispersion due to the blocking elements can be eliminated by a suitable choice of parameters in an L-shaped filter.

When type 1A-501 high-frequency switching diodes are used in the very short wavelength range, it is possible to neglect the stray capacitance of the diode housing and the inductance of the p — n junction without noticeable distortion of the phase shifter's frequency response. A good match can be provided by making the phase shifter design coaxial.

As has been previously shown, for discrete time delay settings with a discrete step τ the binary arrangement is the most efficient way of minimizing the number of diodes used in a phase shifter (Fig. 1).

The phase shifter is composed of k stages which provide $M - 2^k$ discrete delay values, including $\tau_0 = 0$. Each stage includes three switching diodes that connect and disconnect the specified delay τ_n, and blocking elements. The value of the delay τ_n for a particular stage n is made equal to $\tau \cdot 2^{n-1}$, where τ is the specified discrete step. The amount of delay inserted τ_m, where m can assume the values 0, 1, 2, . . . , M − 1, is equal to the sum of the connected delays.

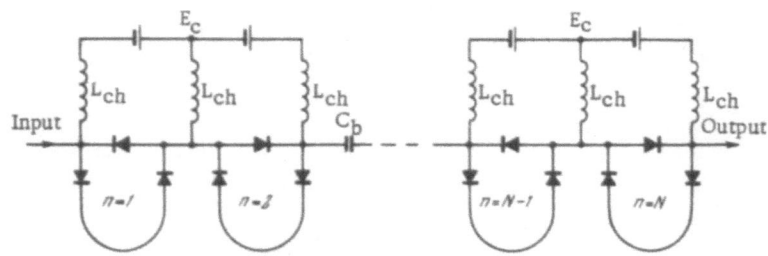

Fig. 1. Binary phase shifter circuit.

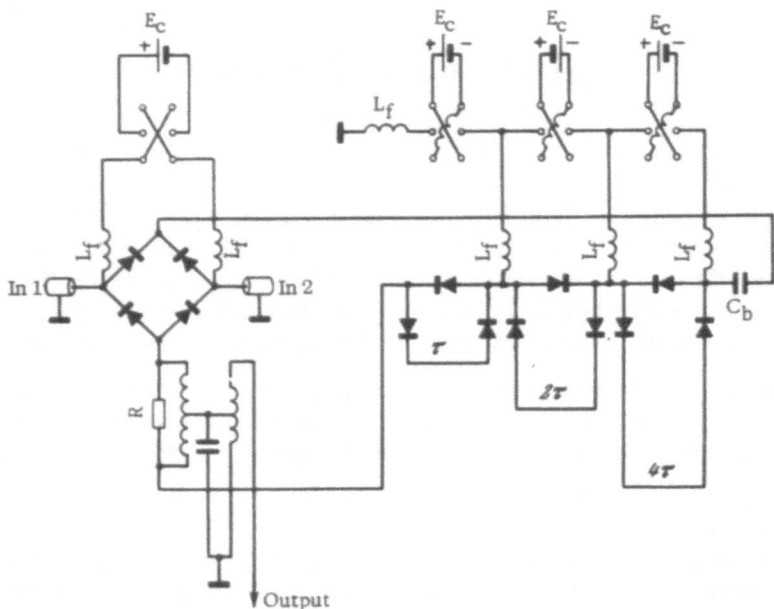

Fig. 2. Schematic circuit of a discrete phase shifter.

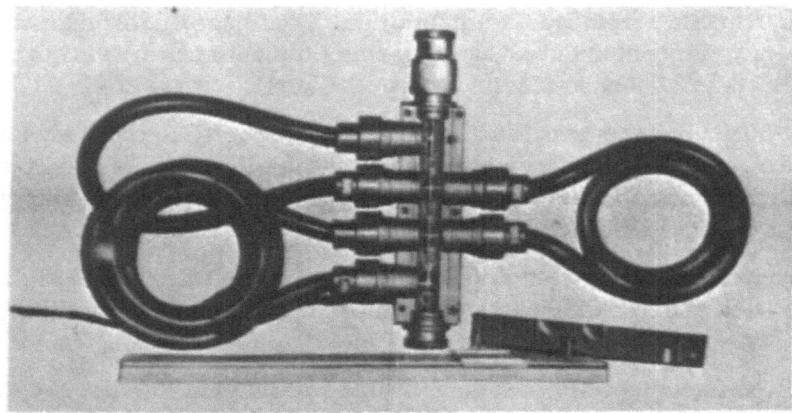

Fig. 3. Construction of the three-stage phase shifter.

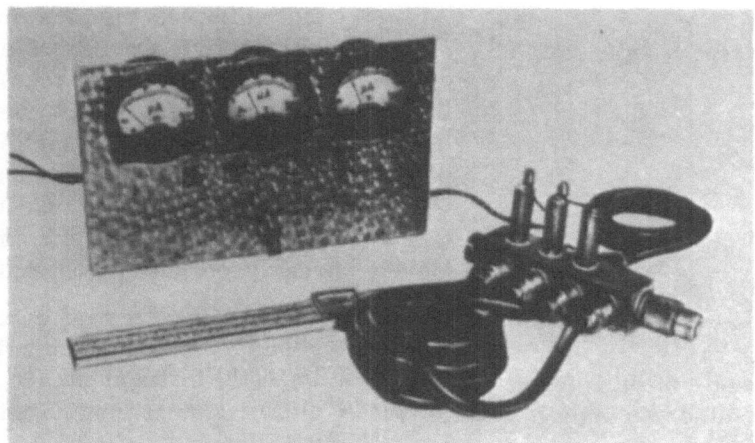

Fig. 4. Control unit and monitor of the phase shifter.

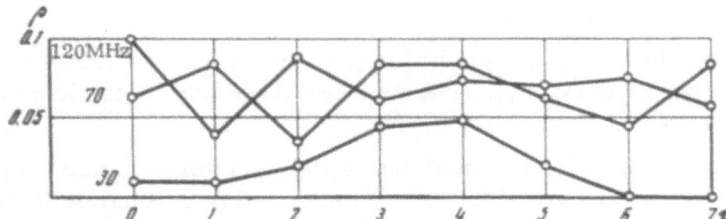

Fig. 5. Reflection coefficient ρ as a function of the discrete steps τ for various frequencies (as indicated by the numbers).

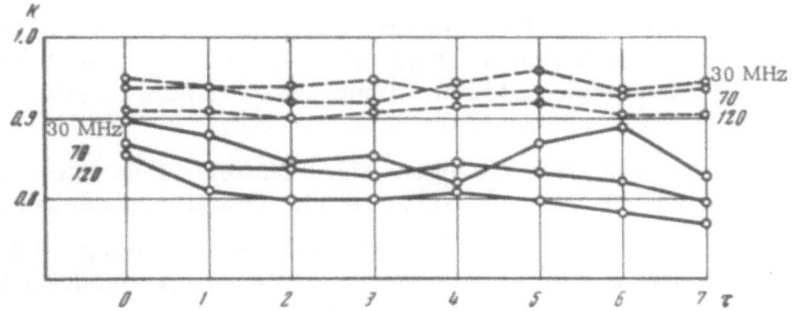

Fig. 6. Transfer constant (K) as a function of the discrete steps τ for various frequencies at input 1 (dashed lines) and input 2 (solid lines).

A schematic circuit of the phase shifter for the DKR-1000 north—south antenna is shown in Fig. 2. The differential adder is similar to the circuit in [3]. It combines the signals from two antennas, one of which is delayed by a time τ_m. The switch connects the delay into the required branch of the differential adder. The phase shifter has three stages with eight discrete values of delay from m = 0 to m = 7. The delays are obtained from type RK-103 cable in discrete steps of delay $\tau = l/c$, where l is the length of the delay cable in centimeters with the curtative coefficient taken into account, and c is the velocity of light.

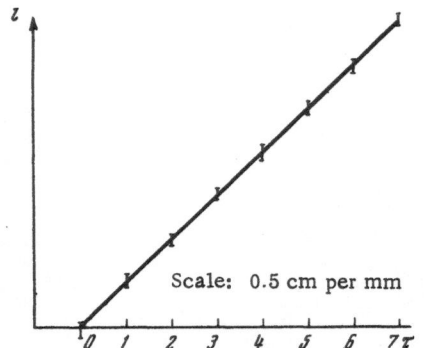

Fig. 7. Electrical length of the cable l
as a function of the discrete steps τ.

Control voltage is supplied individually to each stage, unequivocally assuring the connection or disconnection delay according to the polarity of the control voltage on the stage. Type 1A-501E diodes are used. The diode current is supplied at 50 mA, thus providing an optimum operating condition for the diode. The blocking elements are so chosen (L_f = 43 H and C = 5000 pF) so that they do not affect the high-frequency energy propagation constant.

The coaxial construction of the phase shifter is advantageous for both the mounting and the hermetic sealing of the instrument. An overall view of the phase shifter and its control unit is shown in Figs. 3 and 4.

2. Experimental Results

The discrete diode phase shifter was experimentally checked over the frequency range from 30 to 120 MHz for matching at inputs 1 and 2 (Fig. 2).

As may be seen in Fig. 5, the phase shifter provides a good match over the entire frequency range for all values of delay τ_m. The average value for the modulus of the reflection coefficient $|\rho|_{av}$ = 0.07.

Figure 6 gives the function for the modulus of the transfer constant $|K|$ = F(f MHz, τ_m). From the value of $|K|$ at a frequency of 30 MHz with τ_0 = 0 it is possible to estimate the transfer constant of one diode which is approximately $|K_d|$ = 0.96-0.97. The periodicity in the variation of $|K|$ = F(τ_m) is due to the connection of a different number of diodes for the various values of delay τ_m. The reduction of $|K|$ as the frequency is increased is the result of attenuation in the delay cable. The difference between the transfer constants for inputs 1 and 2 is due to the absence of diodes in one of the branches.

A measurement of the phase shifter's electrical length as a function of the connected delay τ_m (Fig. 7) shows that the deviation of the measured value of τ_m from the specified value is less than 1-2 cm. This corresponds to a maximum phase dispersion at the 120-MHz frequency of no more than $\pm5°$ with a standard deviation of $\pm1.7°$.

It should be noted that measurements of the phase shifter's parameters made with diodes of other types, particularly types D11 and D311, gave substantially poorer results.

The transfer constant for type D11 diodes was substantially lower: $|K_d|$ = 0.9. With type D311 diodes the phase shifter's parameters showed a marked dependence on frequency.

It might be well to point out in conclusion that by virtue of its good high-frequency characteristics, the discrete type of diode phase shifter can be operated at very short wavelengths.

LITERATURE CITED

1. S. N. Ivanov, N. A. Penin, N. E. Skvortsova, and Yu. F. Sokolov, Physical Operating Principles of Superhigh Frequency Semiconductor Diodes, Sovetskoe Radio (1965).
2. Yu. P. Ilyasov and S. N. Ivanov, in: Wideband Cruciform Radio Telescope Research (D.V. Skobel'tsyn, ed.), Consultants Bureau, New York (1969), p. 13.
3. C. L. Ruthroff, Proc. IRE, 47:1337 (1959).

RADIOMETERS WITH 1.6- AND 3.3-cm BAND
PARAMETRIC AMPLIFIERS FOR THE RT-22 TELESCOPE

V. P. Bibinova, A. D. Kuz'min,
M. T. Levchenko, V. I. Pushkarev,
A. E. Salomonovich, and I. V. Shavlovskii*

Radio-astronomical investigations of the planets, discrete sources, and other subjects at centimeter wavelengths require radiometric instrumentation of ever-increasing sensitivity.

As is well known (see [1]), one of the most effective ways of improving radiometer sensitivity is to reduce the receiver noise temperature. Additional sensitivity can be obtained by extending the pre-detector bandwidth of the radiometer. In this connection, a program has been started in 1961 at the FIAN Radio Astronomy Station on the improvement of the sensitivity of modulation-type radiometers operating in the 3- and 1.6-cm waveband for the RT-22 radioscope by using low-noise parametric amplifiers and a wideband intermediate-frequency amplifier.

1. Radiometer Circuits

The 3-cm radiometer block diagram is shown in Fig. 1. The radiometer uses a reflex parametric amplifier with a semiconductor diode and a single resonant circuit. To ensure a quasi-degenerate and coherent amplifier operating mode, pumping and heterodyning is accomplished with a single klystron, the pumping frequency f_p, equal to twice the local oscillator frequency $2f_h$, being provided by a frequency doubler.

Under such conditions, amplification of a noise signal, with a constant spectral density, within the amplifier bandwidth takes place at a minimum noise factor in two sidebands of the paramagnetic amplifier resonance characteristic, which are located symmetrically with respect to the local oscillator center frequency. To stabilize the local oscillator and pumping frequencies, the klystron frequency is passively stabilized by pulling it by means of a high-Q reference resonator. The stabilization factor is 15-20 with a power loss of 3-4 dB in the resonator.

The 1.6-cm radiometer differs from the 3-cm radiometer in that a separate klystron oscillator is used for pumping the parametric amplifier. The frequency of this oscillator is also stabilized by an external resonator.

The two parametric amplifiers are designed in the form of two "lap" joined waveguides. The parametric diode is mounted along the line of intersection of the waveguide symmetry

---

*Deceased.

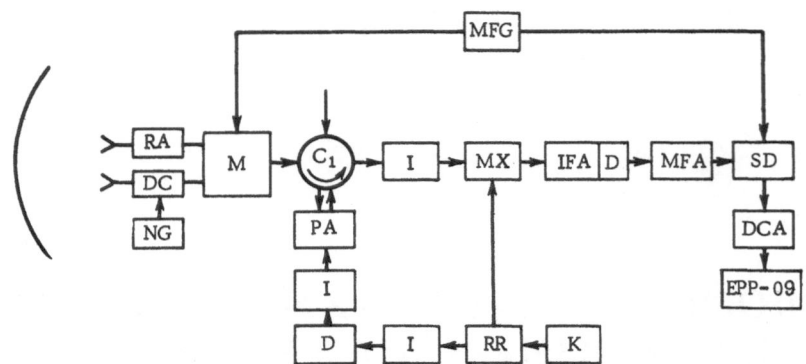

Fig. 1. Block diagram of the three-centimeter band radiometer.
DC) Directional coupler; NG) reference noise generator; RA) remote-
controlled attenuator for setting the quasi-null operating mode; M)
modulator; C) circulator; PA) parametric amplifier; I) isolator; D)
frequency doubler; RR) reference resonator for klystron frequency
stabilization; K) klystron; MX) mixer; IFA) intermediate-frequency
amplifier; D) detector; MFG) modulation-frequency generator;
MFA) modulation-frequency amplifier; SD) synchronous detector;
DCA) direct-current amplifier.

planes and for tuning can be moved out of the signal waveguide into the pump waveguide. The
free ends of the waveguides are short-circuited by plungers which also provide an effective
means for tuning both the pump and signal channels. The parametric amplifier and the frequency doubler diodes operate with self-bias. The parametric-diode frequency doubler has an
efficiency of 15% and an output power of 12-15 mW.

To reduce mutual interference, the parametric amplifier, frequency doubler, and reference resonator are decoupled by ferrite isolators. Particular attention has been paid to preventing the local oscillator signal from leaking into the radiometer input. To achieve this a ferrite
isolator providing a decoupling of 60 dB was placed in front of the mixer.

The high-frequency unit of the radiometer is placed near the principal focal point of the
RT-22 antenna. To reduce the effect of ambient temperature variation, the parametric amplifier,
frequency doubler, and input ferrite devices are separated from the heat-producing units (klystron and intermediate-frequency amplifier) and mounted in a special temperature-stabilized
housing. Dielectric waveguides are used to couple these units to the waveguide channel.

Both radiometers are of the modulation type. Pattern modulation is used to reduce the effect of background radio-frequency radiation from the sky, earth, and surrounding objects; an
auxiliary horn positioned near the antenna focal point serves as a dummy.

The radiometers are calibrated with the aid of gas-discharge noise sources connected
through directional couplers (30 dB) in the dummy channel.

The 3-cm radiometer operates in a quasi-null mode, i.e., with equal noise radiation applied to the modulator input through the signal and dummy channels. The latter is accomplished
by including a special attenuator in the dummy channel. The attenuator is remote controlled for
setting the quasi-null operating mode.

2. Basic Radiometer Characteristics

The effective noise temperature of a receiver using a high-frequency amplifier is given by the well-known expression (see [1]):

$$T_n = (L_c' - 1)\,T_0 + L_c'T_{pa} + \frac{L_c'}{G_{pa}}\,[(L_c'' - 1)\,T_0 + L_c''T_r].$$

Here, L_c' is the loss in the channel between antenna and parametric amplifier, L_c'' is the loss in the channel between the parametric amplifier and mixer, T_0 is the ambient temperature, T_{pa} is the noise temperature of the parametric amplifier, G_{pa} is the parametric amplifier gain, and T_r is the noise temperature of the mixer receiver.

Taking into account that in the two-band operating mode

$$T_r = \tfrac{1}{2}\,(F_r - 1)\,T_0,$$

where the mixer-receiver noise factor is

$$F_r = L(F_{ifa} + t_n - 1),$$

and substituting the following parameters obtained for the 3-cm radiometer:

$$L_c = 1.2\ (0.8\,\text{dB}),\quad L_c'' = 1.3\ (1.1\,\text{dB}),$$
$$T_{pa} = 150°\ \text{K},\quad G_{pa} = 15.8\ (12\ \text{dB}),$$
$$L = 4,\quad F_{ifa} = 3.2,\quad t_n = 1.7,$$

we get $T_n = 425°K$. With an intermediate-frequency amplifier bandwidth $\Delta f = 60$ MHz, this corresponds to a theoretical noise sensitivity $\delta T = 0.09°K$ at $\tau = 1$ sec.

For the 1.6-cm waveband:

$$L_c' = L_c'' = 1.2(0.8\,\text{dB}),\quad T_{pa} = 200°\text{K},$$
$$G_{pa} = 12\ \text{dB},\quad T_r = 3000°\text{K},$$

so that $T_n = 850°K$. For $\Delta f = 56$ MHz, this corresponds to a theoretical noise sensitivity $\delta T = 0.2°K$ at $\tau = 1$ sec.

The measured noise sensitivity of the radiometers has been found as 0.12 and 0.3°K for $\tau = 1$ sec at $\lambda = 3.3$ and 1.6 cm, respectively, i.e., somewhat less than the theoretical values.

Prolonged operation (since 1962) proved that the 3-cm radiometer operates reliably under various weather and seasonal conditions. Relative changes in the radiometer gain in the case of prolonged (6-8 h) continuous operation were on the average 10^{-2} per operating hour. The overall operating time of the radiometer between preventive maintenance checkups and tune-ups was about 1500 h.

The use of parametric amplifier radiometers with the RT-22 radio telescope allowed systematic observations of radio emission from Venus [2, 3] and some other discrete sources, including those exhibiting an anomalous frequency dependence of their flux density (sources 3C84, 3C273, 3C279) [4]. The described radiometer has been used also for determining the precise polarization characteristics of the Taurus A source [5].

High sensitivity of the radiometer made it possible to employ the radio-guidance method using the 3-cm signal of discrete sources, with simultaneous analysis of their radio emission in the millimeter waveband. The radio emissions of the Orion and Omega nebulae [6] and the radio brightness of the Taurus A source were measured in this manner. The 3-cm radiometer served as a basis for the design of a spectral radiometer which helped to detect the spectral emission lines of excited hydrogen [7].

The 1.6-cm radiometer is less reliable but still enabled us to obtain valuable data on radio emission from Venus [3] and other discrete sources.

LITERATURE CITED

1. A. D. Kuz'min and A. E. Salomonovich, Radio Astronomy Methods of Measuring Antenna Parameters [in Russian], Izd. Sovetskoe Radio (1964).
2. V. P. Bibinova, A. D. Kuz'min. A. E. Salomonovich, and I. V. Shavlovskii, Astron. Zh., 39:1083 (1962).
3. Yu. N. Vetukhnovskaya, A. D. Kuz'min, B. G. Kutuza, B. Ya. Losovskii, and A. E. Salomonovich, Izv. Vuz. Radiofiz., 6:1054 (1963).
4. A. Kh. Barret, V. G. Kutuza, L. I. Matveenko, and A. E. Salomonovich, Astron. Zh., 42:527 (1965).
5. V. A. Udal'tsov, Dissertation, FIAN (1966).
6. A. D. Kuz'min and A. E. Salomonovich, Astron. Tsirk., No. 260 (1964).
7. R. L. Sorochenko and E. V. Borodzich, Dokl. Akad. Nauk SSSR, 163:603 (1965).

PARAMETRIC AMPLIFIER FOR RADIO-ASTRONOMICAL INVESTIGATIONS AT 21 cm

I. I. Berulis, B. Z. Kanevskii,
E. A. Spangenberg, and I. A. Strukov

Advances in the development of spectral investigations of cosmic radio emission are determined to a large extent by the possibilities of increasing the sensitivity of radio-astronomical equipment. One of the basic methods of increasing the sensitivity is the use of low-noise parametric amplifiers PA for preamplification at high frequencies. Despite the fact that quan-

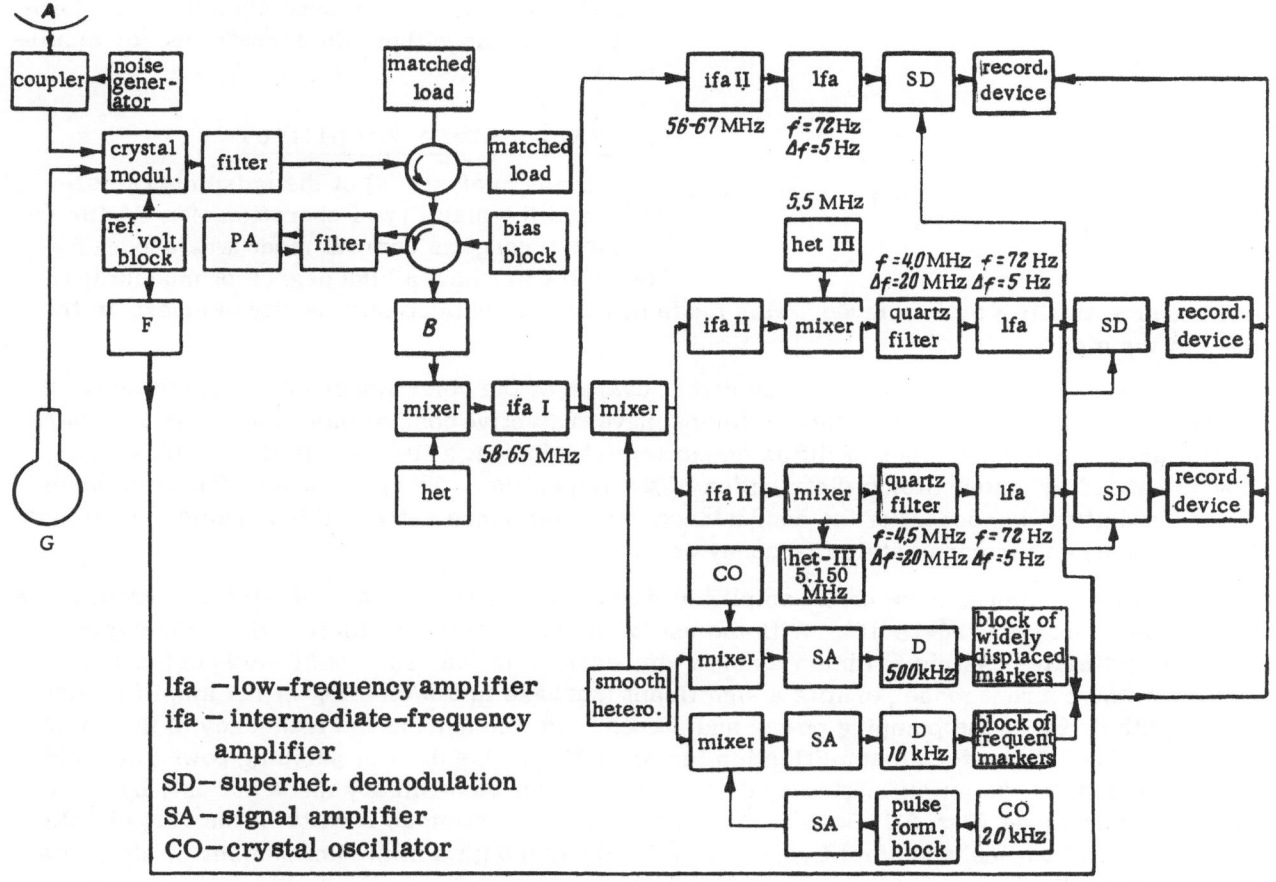

Fig. 1. Block diagram of the radiometer with the parametric amplifier.

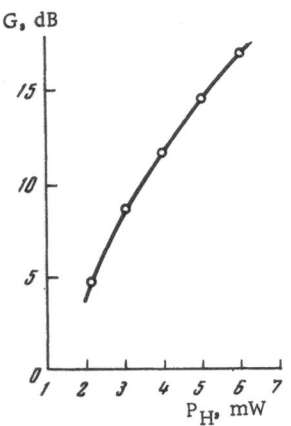

Fig. 2. Amplification factor as a function of the pumping power.

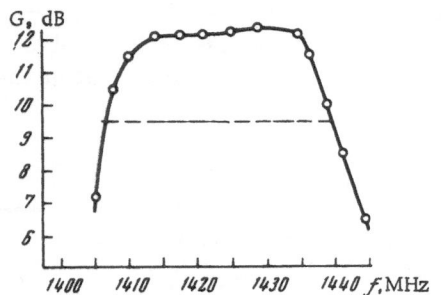

Fig. 3. Amplitude — frequency characteristic of the PA.

tum paramagnetic amplifiers ensure a lower noise temperature and are successfully used for observations at 21 cm [1, 2], the complexity of their operation in some cases makes it preferable (for long-term programs of observations of the distribution of neutral hydrogen in the galaxy) to use PA.

For this purpose, a PA was installed in the 21-cm spectral radiometer of the 22-m radio telescope of the Physics Institute of the Academy of Sciences of the USSR [3] in the beginning of 1966. The overall block diagram of the spectral radiometer with the PA is presented in Fig. 1. It is a modulation-type superheterodyne receiver with triple frequency conversion and smooth variation of the frequency in the second heterodyne. The modulation is accomplished by switching the input of the receiver from the antenna to an equivalent with the effective temperatures of the two equalized before the observations. In the radiometer the PA is connected to the circuit in the reflex scheme with the use of two ferrite circulators having losses in one arm equal to 0.3 dB. The remaining part of the radiometer is the radiospectrometer prepared for observations without the high-frequency amplifier [4].

1. Parametric Amplifier

In the analysis [5] of the noise characteristics and the stability of operation of a PA for the decimeter range, a current mode was used for a 21-cm PA in which a high degree of modulation and better stability were achieved. This mode utilizes the buildup of minority carriers in the base of the diode.

An examination of pumping in the direct branch of the volt—ampere characteristic of the diode showed a diffusion conductance with capacitive and active components, along with the charge capacitance. In certain types of diffusion parametric diodes, a field exists in the diode base that is caused by a nonuniform distribution of the impurities. The presence of this field leads to the capacitive component of the conductance becoming more active, which enables one to use this conductance for parametric frequency conversion.

The noise temperature of the amplifier, for the operation of the diode in the current mode, is 100–150°K (for G = 10–15 dB). With the use of diodes operating with negative bias voltage, it is theoretically possible to obtain a noise temperature of the order of 30–50°K in the PA. However, the current mode permits a significant increase in the stability of the amplification factor with drifts of the pumping power and, hence, with the drift of the frequency of the pumping oscillator. Thus, for an amplification factor of 15 dB, the drift of pumping power by 1 dB leads to a drift of the amplification by 6 dB in the case of the ordinary mode but only by 2 dB in the current mode (Fig. 2). Besides, the use of the current mode results in the very simple construction of a wideband amplifier (10–15% band pass) with a minimum number of elements needing fine adjustment.

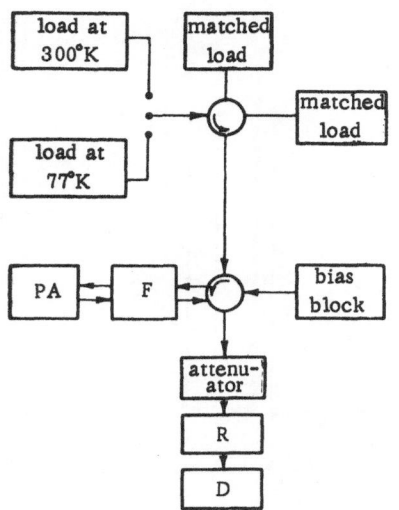

Fig. 4. Block diagram of the equipment for measuring noise temperature of the PA.

A coaxial stub of length $\lambda/4$ corresponding to the no-load frequency is placed in the amplifier after the diode; the stub represents an inductance at the signal frequency. The pumping power is fed to the diode through a waveguide formed by the extension of grounded plates. The pumping waveguide is beyond critical for the no-load frequency. The no-load circuit is formed only by the elements of the frame of the diode, since in the current mode the capacitance of the $p-n$ junction C_0 is appreciably larger than the capacitance of the diode socket; therefore, the resonant frequency of the no-load circuit is determined mainly by the parasitic parameters L_{bb} and C_n of the diode frame. The no-load circuit can be tuned over small ranges by including a small fine-tuning capacitor in parallel with the capacitance of the frame.

The signal circuit is formed by the capacitance of the $p-n$ junction, the inductances of the stub after the diode, and the segment of the high-frequency line included between the diode and the collectors. The main characteristics of the PA are: input frequency f_s, 1420, 405 MHz; capacitance of the $p-n$ junction C_{bb}, 0.22 ± 0.04 nF; noise temperature T_n, 120° K; amplification factor G_{PA}, 12 dB; and consumable pumping power P_P, ≤ 5 mW.

The amplitude—frequency characteristic of the PA is shown in Fig. 3. It should be mentioned that because of the narrow band characteristics of the decoupling devices the PA was specially tuned to a small band. The use of the current mode in the decimeter range enables one to obtain a 10–15% band pass without additional compensating circuits in the presence of wide-band couplings.

2. Noise Measurements

The sensitivity of a radiometer is mainly determined by the fluctuation of internal noises and the instability of the amplification factor. The fluctuational sensitivity caused by internal noises, is determined by the relation

$$\Delta T_{\min} \approx \frac{\pi}{2} \frac{T_n}{\sqrt{\Delta f \tau}}, \tag{1}$$

where T_n is the overall noise temperature of the system, Δf is the band pass of the receiver, and τ is the time constant.

In the presence of fluctuations of the amplification factor G, the sensitivity is lowered by an amount determined by the relation

$$\Delta T_G = (\gamma - 1)(|T_a - T_e|), \tag{2}$$

where $\gamma = 1 + (\Delta G/G)$ is the coefficient of fluctuations of amplification, T_a is the antenna noise temperature, and T_e is the noise temperature of the equivalent.

A well-matched load cooled to the temperature of liquid nitrogen was used as the equivalent. The noise temperature of the equivalent together with the connecting feeder, was 90° K; according to the measurements made, T_a was 43° K. In order to obtain maximum sensitivity [according to (2) it is necessary to equalize T_a and T_e] additional noises were introduced in the channel. The noise temperatures of the antenna and the equivalent were successfully balanced

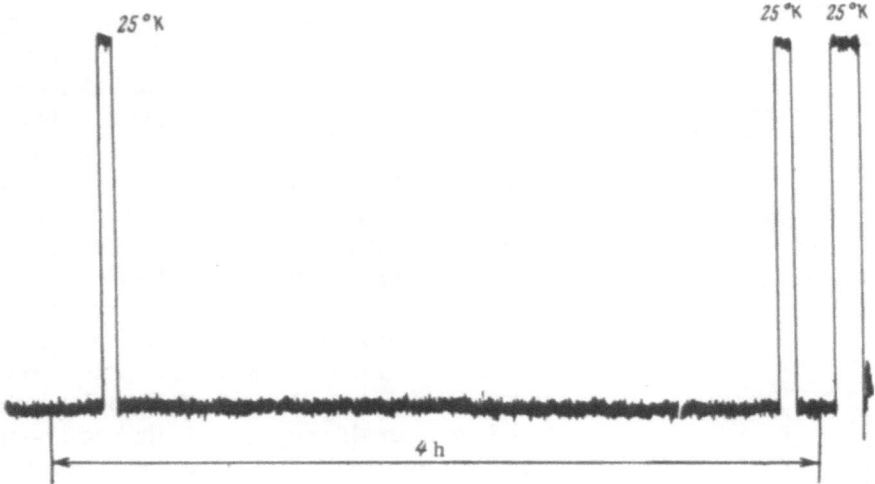

Fig. 5. A record characterizing the stability of the radiometer with the
PA in continuous spectrum (time constant, 1 sec).

to within 1.5°K. As a result, the actual sensitivity of the radiometer can be written as

$$\Delta T_0 = \Delta T_G + \Delta T_{\min}.$$ (3)

The overall noise temperature of the entire radiometer system with the PA is determined by
the expression

$$T_{\mathrm{n}} = T_{\mathrm{a}} + \frac{1}{q}\left[T_0(1-q) + T_{\mathrm{PA}} + \frac{T_{\mathrm{r}}}{G_{\mathrm{PA}}}\right],$$ (4)

where T_{PA} is the noise temperature of the PA; T_{r} is the noise temperature of the radiometer
without the PA; q is the transfer coefficient of the channel up to the PA; and G_{PA} is the amplification factor of the PA.

The noise temperature of the radiometer with the PA was measured with the use of two
noise equivalents consisting of well-matched loads [voltage standing wave ratio (vswr) ≤ 1.04];
one of these was kept at room temperature ($T_{\mathrm{room}} = 273°K + t°C$), the other was cooled to the
temperature of liquid nitrogen ($T_{\mathrm{cool}} = 77.3°K$). The block diagram of the arrangement for the
noise measurements is presented in Fig. 4.

The noise powers from the two loads were alternately connected to the circulator system
and were amplified by the PA and the second cascade of amplification. A radiometer with the
noise temperature $T_{\mathrm{r}} = 1500°K$ was used as the second cascade. A precision attenuator connected before the detector made it possible to obtain the same level of power at the output of
the receiver.

The difference of the attenuator readings, corresponding to the ratio of the noise powers N,
is related to the total noise temperature of the two amplifiers connected in series, by the simple
relation

$$T_{\mathrm{n.s}} = \frac{T_{\mathrm{room}} - NT_{\mathrm{cool}}}{N-1},$$ (5)

where

$$T_{\mathrm{n.s}} = T_{\mathrm{PA}} + \frac{T_{\mathrm{r}}}{G_{\mathrm{PA}}}.$$

Equation (5) is conveniently represented in the form

$$T_{\mathrm{PA}} = \frac{T_{\mathrm{room}} - NT_{\mathrm{cool}}}{N-1} - \frac{T_{\mathrm{r}}}{G_{\mathrm{PA}}}.$$ (6)

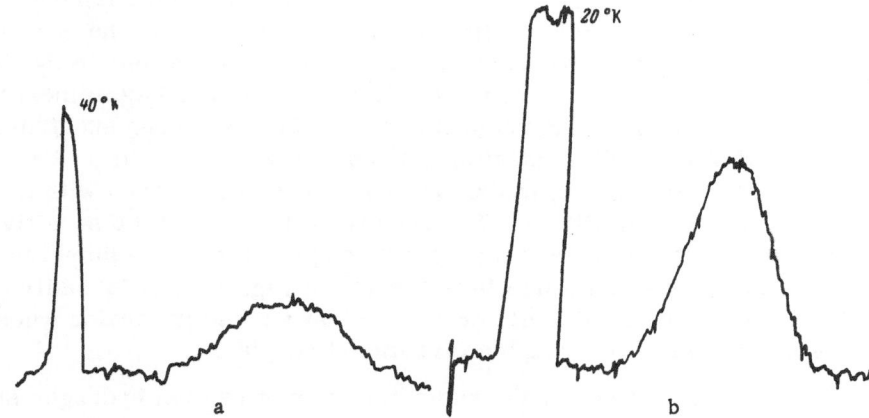

Fig. 6. Record of the radio source Virgo A in the continuous spectrum
(a) with and (b) without the PA.

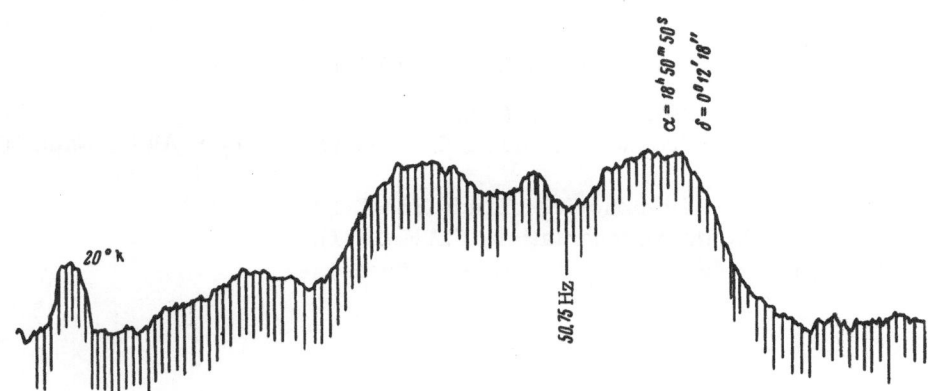

Fig. 7. Record of the profile of the emission line of neutral hydrogen
(frequency marks at 10-kHz intervals).

As a result of successive measurements the noise temperature of the PA was determined for different values of its amplification factor. The optimum parameters of the PA, for which ΔT_0 becomes minimum, were found to be as follows: T_{PA} = 120°K, G_{PA} = 12 dB, and Δf = 32 MHz. In the measurements of T_{PA} and G_{PA}, the linearity of the system after the PA was checked. Also, in view of the use of wideband loads, the entire system was tested for interference stability. The transfer factor of the high-frequency circuit up to the PA was determined by the losses in the modulator, circulators, and the connecting feeder lines (q = 0.82). Substituting the values of T_a, q, $T_{PA.}$, and T_r/G_{PA} in Eq. (4), we get T_n = 420°K, which is in good agreement with the measured data. The total amount of error in the measurements did not exceed 5°K.

The long-period stability of the amplifier was investigated before carrying out observations. A typical record of the output signal during a period of 4 h is presented in Fig. 5 for an amplification factor of 12 dB.

From a large number of measurements of the stability of the PA at G_{PA} = 12 dB the maximum drift of the amplification factor over a long period (of the order of 8 h) was found to be 0.5 dB. The main source of the slow instabilities was the variation in the pumping power. The frequency of the pumping power was stabilized.

The PA was not regulated in the process of operation. At a noise temperature of 420°K for the entire system the fluctuational sensitivity for measurements in the continuous spectrum (τ = 1 sec and Δf = 7 MHz) corresponded to 0.25°K and for observations in the radio line (τ = 30 sec, Δf = 20 kHz), it was equal to ΔT_0 = 0.84°K, which is in good agreement with the computed data. Examples of records in the radio line and in the continuous spectrum are shown in Figs. 6 and 7. A record of the radio emission of the discrete source Virgo A with coordinates $\alpha = 12^h 29^m 09^s$, $\delta = 12°34'32''$ (1966.5) in the continuous spectrum, taken with and without the parametric amplifier, is shown in Fig. 6. It is seen from Fig. 6 that the sensitivity of the radiometer with the PA increased by a factor of 3.6 compared to the radiometer without the PA. An example of the record of the radio line is presented in Fig. 7. The intensity of the received radio emission in the observations was calibrated against a noise generator whose power was introduced into the antenna circuit through a directional coupler.

The preliminary investigations of the radio emission of neutral hydrogen showed that the radiometer has a high sensitivity in the narrow band of reception and permits the separation of fine details in the form of lines, which had not been noted earlier; it also makes it possible to study in detail the distribution of hydrogen in the galaxy.

The authors express their gratitude to R. L. Sorochenko and V. S. Etkin for discussion of the obtained results.

LITERATURE CITED

1. T. V. Jelly, Microwave J., Vol. 5, No. 2 (1962).
2. R. M. Martirosyan, A. M. Prokhorov, and R. S. Sorochenko, Dokl. Akad. Nauk SSSR, 156: 1326 (1964).
3. P. D. Kalachev and A. E. Salomonovich, Trudy FIAN, 17: 11 (1962).
4. R. S. Sorochenko, Candidate's dissertation, FIAN (1961).
5. K. S. Mosayan, A. I. Strukov, and V. S. Etkin (in press).

DETERMINATION OF THE ORIENTATION OF THE ELECTRIC AXIS FOR THE EAST—WEST ARRAY OF THE DKR-1000 CRUCIFORM RADIO TELESCOPE BY STATISTICAL REDUCTION OF OBSERVATIONS OF MANY DISCRETE SOURCES

V. V. Vitkevich and V. N. Kozhukhov

1. Developments in radio astronomy, and in particular the discovery of numerous discrete radio sources in the sky, have opened up new potentialities for investigating antenna systems. If the parameters of a source are known, it becomes possible to determine the effective area of an antenna and to study radio reception diagrams, including the magnitude and direction of the side lobes, by utilizing discrete sources as natural "radio transmitters" over a wide frequency range. As the antenna size increases and the far-field pattern becomes located at greater distances, this method becomes even more advantageous, because local oscillators must be placed at an excessive distance if they are used to plot the patterns of large antennas, rendering the work much more difficult and expensive.

Until recently the method most often employed was based on the use of individual, well-studied discrete sources of maximum intensity (see, for example, Kuz'min and Salomonovich [1]). A difficulty arises here in that we cannot change the coordinates of the sources; they are not always located where they are needed, and they are few in number.

An alternative possibility would be to use a large number of sources for alignment purposes and to derive the mean values of the required parameters from observations. The inadequacy of the information available on each source involved in such a program (the low accuracy in the determination of the flux densities and in establishing the coordinates and angular diameters) is compensated by the large number of sources, enabling an average to be taken over a substantial number of observations.

We have taken advantage of this possibility to determine the orientation of the east — west array of the wide-band DKR-1000 cruciform radio telescope. This array forms a parabolic cylinder measuring 40 × 1008 m with capability for rotation in altitude [2]. According to geodetic data its azimuth agrees accurately with the south direction, but its west side has been raised by a 30-min angle above its east side due to terrain conditions.

The basic material for securing the necessary data was the first series of observations carried out for the purpose of compiling a catalog of discrete radio sources at the 3.5-m wavelength [3].

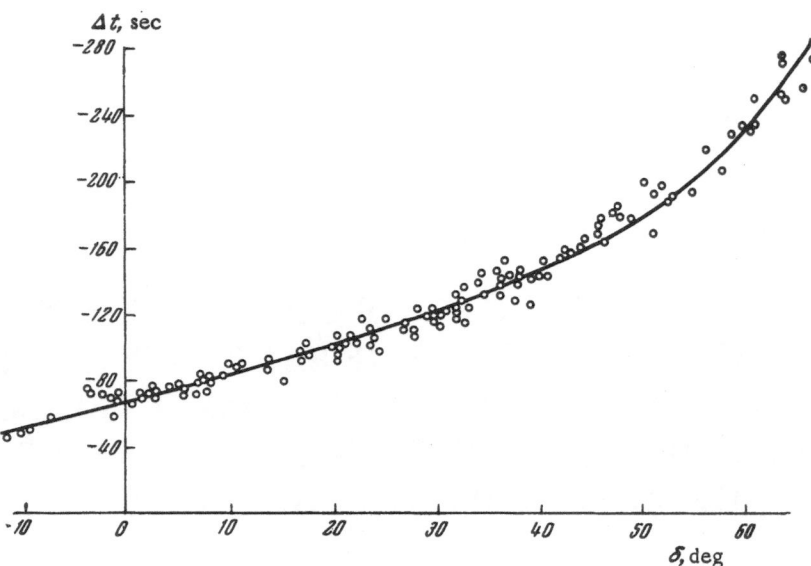

Fig. 1. Difference between the actual transit time of radio sources
through the antenna beam and their right ascension ($\Delta t = s - \alpha$) as
a function of the declination δ. The symbols designate the sources
from which the orientation of the east — west array of the DKR-1000
was determined; the solid curve is based on the computed parameters.

2. The observational data for each source were used to determine the sidereal time s of
its culmination; the quantity $\Delta t = s - \alpha$ was found (α is the right ascension) and a graph was
drawn for the dependence of Δt on the declination δ. If errors of all kinds were absent, Δt
would be equal to zero; that is, all the sources would simply fall along a line in the graph coin-
ciding with the axis of abscissas.

Figure 1 illustrates the results of the reduction that was made. It differs from the ideal
case we have described in two respects. In the first place, the curves do not coincide with the
axis of abscissas (that is, $\Delta t \neq 0$); there is a systematic deviation, depending on declination,
because the east — west array of the DKR-1000 is not oriented exactly along the east — west
line, and moreover is inclined. Secondly, the points exhibit a certain amount of scatter in the
graph. Two groups of errors are responsible for the scatter. The first group includes systematic errors relative to a given source, that is, possible errors in the coordinates of the source,
and the effect of confusion of the main lobe resulting from the presence of another more or less
strong source in the immediate vicinity of the source under study. When many sources are observed, however, these errors make no systematic contribution to the results. The second
group consists of purely random errors, which should disappear if a sufficiently large number
of records are averaged. These include errors of measurement, the influence of the ionosphere,
the effects of confusion from the side lobes (as will occur if a strong source falls within one of
the side lobes and if the lobe is unstable), and mismatch in the phases of the antenna arrays.

The rather extensive observational material obtained through the analysis has enabled
the inclination and azimuth of the east — west array of the DKR-1000 to be determined. The influence of the inclination on the transit time of a source is expressed by the relation

$$(\Delta t_1) = \frac{a \cos z}{\cos \delta} ,$$
(1)

where z is the zenith distance and δ is the declination of the source; a represents the inclination of the array expressed in the same units as the quantity Δt.

Correspondingly, the influence of the azimuth is expressed by the relation

$$(\Delta t)_2 = \frac{b \sin z}{\cos \delta} , \qquad (2)$$

where b is the deviation of the east — west array from the true east — west direction. Combining Eqs. (1) and (2), we have

$$\Delta t = (\Delta t)_1 + (\Delta t)_2 = \frac{a \cos z}{\cos \delta} + \frac{b \sin z}{\cos \delta}$$

or

$$a + b \tan z = \frac{\cos \delta}{\cos z} \Delta t. \qquad (3)$$

The method of least squares may then be used to determine the values of a and b of interest to us. For this purpose we have utilized 131 sources observed during the period from March to May, 1965. The results of the computations have yielded the following values:

$$a = 29'10'' \pm 34'' \text{ (west end higher),}$$
$$b = 24'' \pm 43'' \text{ (west end deflected toward south).}$$

The curve shown in Fig. 1 is a reference curve for computing the values of a and b.

The deviation of the observational data from the theoretical curve amounts to $\leq 10^s$. In view of the fact that the values of Δt fluctuate from day to day for the same source over a range of about 30^s while the maximum number of records averaged for a single source is 10, the rms deviation on the diagram should be about 10^s; thus all the sources fall within the range of random error. Consequently, there is no reason to believe that the first group of errors (inaccuracies in the coordinates and the confusion effect) are of significance at wavelength $\lambda = 3.5$ m for the sources we have used.

LITERATURE CITED

1. A. D. Kuz'min and A. E. Salomonovich, Radio Astronomy Methods of Measuring Antenna Parameters [in Russian], Izd. Sovetskoe Radio (1964).
2. V. V. Vitkevich and P. D. Kalachev, in: Radio Telescopes (D. V. Skobel'tsyn, ed.), Consultants Bureau, New York (1966), p. 1.
3. V. S. Artyukh, V. V. Vitkevich, and R. D. Dagkesamanskii, Soviet Astron. — AJ 11:792 (1968).

RADIO INTERFEROMETRY WITH RADIO-RELAY LINKS

V. V. Balinov and V. V. Vitkevich

Introduction

Instrumental resolution has been a primary consideration since the first steps taken in the development of radio astronomy. Soon after the first radio-frequency observations, it became clear that a simple increase in the size of the radio telescopes was not feasible. Radio interferometry appeared. Later, methods were developed for studying the radio brightness of sources during lunar occultation. In recent years, there have been studies of the size of radio sources by means of the flickering due to inhomogeneities in the interplanetary plasma [1].

One of the most important of these approaches has been that of radio interferometry. This method has constantly been improved during the whole development of radio astronomy, and has been used successfully in the USSR and in other countries. In particular, during the initial developmental period of radio astronomy, it was used to a large extent in the Lebedev Physics Institute for various studies [2, 3].

During the initial discussions of the design of a cruciform radio telescope, attention was called to its possible use as one antenna of a radio interferometer [4]. However, simple calculations showed that the design of radio interferometers with signal interference ran into great difficulties at base lines of a few kilometers or more, because of energy channeling losses along the cables. It therefore became necessary to design the radio interferometer with radio-relay links, i.e., with radio (wireless) transmission of the signals.

To our knowledge, the first such radio interferometer was constructed in England at the radio-astronomy observatory at Jodrell Bank [5]. Its basic circuit consists of two radio-frequency antennas connected to the summation point by radio links. The summation point can be at one of the antennas.

One of the simplest systems would seem to be that in which the high-frequency (hf) working signal from the antenna was amplified and transmitted to the summation point. But such a system is essentially useless because the transmitted signal must be thoroughly decoupled from the telescope antenna. A calculation shows, for example, that for meter wavelengths ($\lambda \approx 3$ m), an antenna separation of only 5 km, and a relay antenna of area A = 30 m^2 operating under line-of-sight conditions, the decoupling must be around 10^{-6} in power; in practice, such a decoupling is extremely difficult to achieve. For longer relay distances, this difficulty increases significantly; so this system has not been used. Interferometers with radio-relay links have been used.

We discuss below the general theoretical bases of the operation of an interferometer with radio-relay links and present some data on tests of such an interferometer at the Radio-Astronomy Station of the Lebedev Physics Institute, at Pushchino.

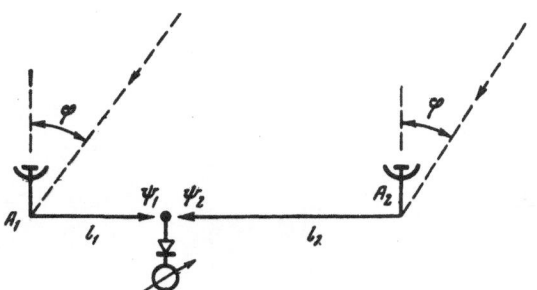

Fig. 1. Interferometer with different path lengths.

1. The Radio Interferometer and the Necessity for Time Delay

Let us consider the interferometer circuit and the parameters which govern its operation (Fig. 1). We denote by ψ_1 and ψ_2 the emf phases at the summation point. The phase difference for the waves coming from A_1 and A_2 is evidently equal to

$$\psi_2 - \psi_1 = kd = 2\pi\left(L_2 - L_1 - \frac{D\sin\varphi}{\lambda}\right),$$

where $d = \gamma(l_2 - l_1) - D\sin\varphi$ is the difference in path lengths taking into account spatial retardation; $\gamma = \lambda/\lambda_c = v/v_c = k_c/k$ is the shortening factor in the cable; and $L_c = l_c/\lambda_c = \gamma l_c/\lambda$ is the electrical length of the cable in units of λ_c.

After a quadratic detector and an integration circuit with a time constant τ, an interference envelope corresponding to the high frequency ω is extracted:

$$P \approx \overline{[\cos\omega t + \cos(\omega t - kd)]^2} = 1 + \cos kd.$$

The lobe maximum (that of the so-called n-th lobe) is given by the obvious relation

$$kd = 2\pi n$$

or

$$L_2 - L_1 - \frac{D\sin\varphi}{\lambda} = n.$$

The lobe width (between maxima and minima) is

$$\Delta\varphi = \frac{\lambda}{D\cos\varphi}.$$

The narrowest lobe thus has a width of $\Delta\varphi = \lambda/D$. This lobe is normal to the interferometer axis ($\varphi = 0$). For a given interferometer, the narrowest lobe has an index

$$n_{\varphi=0} = L_2 - L_1 = \Delta L.$$

An expression was given earlier for $P(\varphi)$ in the case of a monochromatic signal. However, reception always occurs in a finite band $\Delta\omega$, leading to "smearing" of the interference pattern [2]. In this case we have, for a "rectangular" band $\Delta\omega$

$$P \approx 1 + M\cos kd,$$

where

$$M = \frac{\sin\left(\frac{kd}{2}\frac{\Delta\omega}{\omega}\right)}{\frac{kd}{2}\frac{\Delta\omega}{\omega}} = \frac{\sin\left(\pi n\frac{\Delta\omega}{\omega}\right)}{\pi n\frac{\Delta\omega}{\omega}}$$

is the modulation of the interference pattern near the n-th lobe.

It is easy to find the lobe index corresponding to $M = 2/\pi \approx 0.6$:

$$n_{0.6} = 0.5 \, \frac{\omega}{\Delta \omega} \; .$$

Similarly, for a Gaussian band, we have

$$M = e^{-\frac{kd}{2} \frac{\Delta \omega}{\omega}} = e^{-\pi n \frac{\Delta \omega}{\omega}}$$

and

$$n_{0.6} = 0.15 \, \frac{\omega}{\Delta \omega} \; .$$

It follows from these last expressions that the lobe with the maximum modulation of $M = 1$ has the index $n = 0$, i.e., it corresponds to equality of the electrical path lengths in relation to "spatial" path length:

$$L_2 = L_1 + \frac{D \sin \varphi_n}{\lambda} \; .$$

But in order that this lobe be the narrowest, we must have $\varphi = 0$ for it. We can satisfy these two conditions if the following equality holds:

$$L_2 = L_1.$$

This requirement of equal path length also involves the necessity for signal delay (in the first path, if $L_2 > L_1$).

In practice, the whole relay link may be in one of the paths (e.g., in L_2), and thus we would have $L_2 \gg L_1$. To equalize these path lengths, additional delay must be inserted in the L_1 path. If we denote this delay by L_d, we can satisfy the two above conditions if

$$L_d = L_2 - L_1.$$

The quantity L_d is close to L_2 in order of magnitude if the radio link makes the fundamental contribution to L_2. We can evaluate the permissible error in L_d by equating $M = M_{min} = 0.6$, i.e., by setting the modulation of the interference pattern equal to 0.6.

Evidently, we have

$$\Delta L \leqslant \alpha \, \frac{\omega}{\Delta \omega} \; ,$$

and $\alpha = 0.5$ for a rectangular frequency characteristic or $\alpha = 0.15$ for a Gaussian one.

If $\omega / \Delta \omega \approx 10^2$, the permissible error ΔL becomes a few tens of wavelengths.

2. Delay in the Intermediate-Frequency Circuit

Insertion of a delay line meets greater difficulties at high frequencies than at the intermediate frequency. Therefore, we will consider the characteristics of a second type of interferometer (Fig. 2). The phase difference at the summation point is

$$\psi_2 - \psi_1 = kd - k_i l_d = k(d - l_d) + k_b l_d,$$

where

$$d = l_2 - l_1 - D \sin \varphi.$$

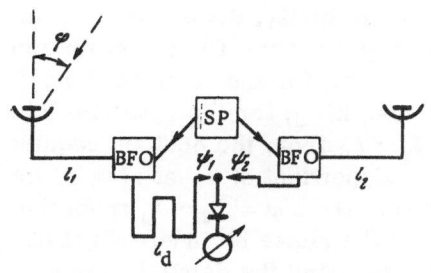

Fig. 2. Interferometer with a delay in the intermediate-frequency circuit.

Here l_d denotes the "electrical equivalent" of the actual length of the delay cable; k, k_b, and k_i are the wave numbers in free space of the working frequency, the beat frequency, and the intermediate frequency; $k_i = k - k_b$.

The power at the input to the integrating circuit for a monochromatic signal is

$$P \approx 1 + \cos{(kd - k_i l_d)}$$

or

$$P \approx 1 + \cos{[k_i (d - l_d) + k_b d]}.$$

For the noise signal in a rectangular frequency band $\Delta\omega$, we have

$$P_{\Delta\omega} = 1 + M \cos{[k (d - l_d) + k_b l_d]},$$

and

$$M = \frac{\sin{\left[\frac{k(d - l_d)}{2} \frac{\Delta\omega}{\omega}\right]}}{\frac{k(d - l_d)}{2} \frac{\Delta\omega}{\omega}}.$$

The position of the n-th lobe (the angle φ_n) is given by

$$kd - k_i l_d = 2\pi n.$$

The width of the lobe "between zeros" is

$$\Delta\varphi = \frac{\lambda}{D \cos{\varphi}}.$$

From these expressions it is easy to find the instability in the lobe control due to the instability of the beat-frequency oscillator (with $\Delta\omega \approx 0$ and for k_i = const):

$$\frac{\delta\varphi (\omega_b)}{\Delta\varphi} = -\frac{d}{\lambda} \frac{\delta\lambda_b}{\lambda_b}.$$

A "minimum" lobe is one of minimum width (for it we always have $\varphi = 0$); a "cophasal" lobe is one which corresponds to equal electrical path lengths to the summation point (for it we always have n = 0); an "optimum" lobe is one for which M is at a maximum (for a point source, we have M = 1); and a "stable" lobe is one whose position is independent of the instability in the beat frequency.

We can determine the conditions under which a given interference lobe would have these properties.

For an interferometer having equal path lengths in the intermediate-frequency circuit, one lobe will be at once cophasal, optimum, and stable. By varying the delay at the high frequency, one can send the lobe in any direction (including the direction $\varphi = 0$, i.e., one can make it minimum).

For an interferometer having a delay in the intermediate-frequency circuit compensating the high-frequency delay time, the pattern changes. The properties listed now will generally

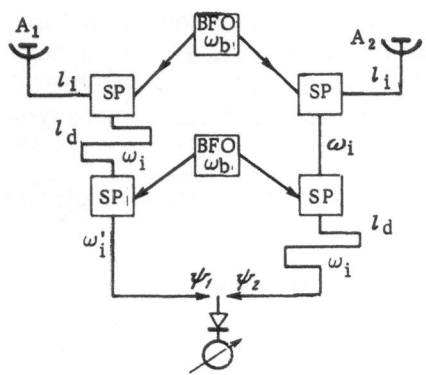

Fig. 3. Interferometer with double frequency conversion with delays l_d and l_d'.

belong to different lobes. Actually, the direction φ is determined in the following manner: for the minimum lobe, we have $\varphi = 0$, as usual; for the cophasal lobe, we have $D \sin \varphi = l_2 - l_1 - (k_i/k) l_d$; for the optimum lobe we find $D \sin \varphi = l_2 - l_1 - l_d$ from the obvious requirement that the phase be independent of k_i at $M = 1$; and for the stable lobe, we find $D \sin \varphi = l_2 - l_1$ from the obvious requirement that the phase be independent of k_b (at $k_i = $ const). We can then find the delay l_d for a given φ: for a cophasal lobe, $l_d = (k/k_i)d$, and for an optimum lobe, $l_d = d$. The magnitude of the delay does not affect the directions of the minimum and stable lobes.

We can draw certain conclusions. Starting from the requirement of an optimum lobe (i.e., for which $M = 1$), we find the delay time in the cable paths should be

$$l_d = d \quad \text{or} \quad \tau_{\mu sec} = \frac{d}{300},$$

and with the auxiliary condition $\varphi = 0$ (for a minimum lobe), we have $d = l_2 - l_1$. Accordingly, it is not the difference in electrical path lengths of the interferometer which must be compensated but the time taken for the signal to travel along them.

The delay in the intermediate-frequency circuit does not affect the stability of the lobe position in space (which is sensitive to the instability of the beat frequency); as before, this stability is given by $\frac{\delta \varphi}{\Delta \varphi} = - \frac{d}{\lambda} \frac{\delta \lambda_b}{\lambda_c}$. However, for greater values of d, upper limits are imposed on $\Delta \lambda_b / \lambda_b$ for an interferometer with radio-relay links. For example, with $d \approx 30$ km and $\lambda \approx 3$ m, we have $\delta \lambda_b / \lambda_b \approx 10^{-6}$ for a permissible instability of $\delta \varphi / \Delta \varphi = 1\%$. It is seen that for this interferometer circuit, the conditions for cophasal and optimum lobes are mutually exclusive.

We describe below a new step in the development of this system, with two converters, for which the two conditions mentioned may coincide.

Figure 3 shows a block diagram of this interferometer. The phase difference at the summation point is

$$\psi_2 - \psi_1 = kd - k_i l_d + k_i' l_d'.$$

The position of the n-th lobe is given by

$$kd - k_i l + k_i' l_d' = 2\pi n$$

or (since $k_i = k_b' + k_i'$ and $k = k_b + k_i = k_b + k_b' + k_i'$)

$$k_i' (d - l_d + l_d') + k_b'(d - l_d) + k_b d = 2\pi n.$$

The conditions for cophasal and optimum lobes follow: for a cophasal lobe, we have $k_i'(d - l_d' + l_d') + k_b'(d - l_d) + k_b d = 0$; for an optimum lobe, we have $d - l_d + l_d = 0$. This system of two unknowns (l_d and l_d') is evidently consistent and has the solutions

$$l_d' = d \frac{k_b}{k_b'}, \quad l_d = l_d' + d.$$

The possibility of a solution of this problem has thus been established.

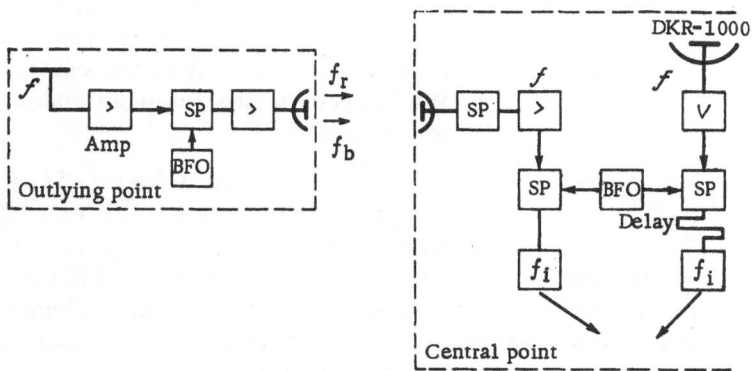

Fig. 4. Block diagram of the radio-interferometer model with radio-relay links used at the Radio-Astronomy Station of the Lebedev Physics Institute.

3. The Radio-Relay Interferometer of the Radio-Astronomy Station of the Lebedev Physics Institute (at Pushchino)

We can discuss some practical aspects of the design of an interferometer with radio-relay links on the basis of an experimental model.

Figure 4 shows a block diagram of the interferometer which was used. The working frequency was f = 86 MHz and the beat frequency was 2000 MHz. The apparatus at the outlying point (at Aladino) consists of an amplifier having a voltage gain of k ≈ 10^7 and a converter (mixer and power oscillator based on GS-90B vacuum tubes) in the transmitting unit of a "Strela-M" radio communications system.

The apparatus at the central point (district DKR-1000) consists of a converter (a double-loop coaxial crystal mixer tuned to the relay frequency f_r of the beat frequency f_b) and a standard radio-astronomy receiver of the compensation or modulation-correlation type. The base line is of the order of 5000 m to the southeast of DKR-1000.

The outlying antenna is a cophasal curtain array with an area A ≈ 10 m^2. The relay antennas are 4-m paraboloids, on a 30-m tower at the outlying point and at a height of 3 m at the central point.

Control recordings were made of sources in 1961 with a similar interferometer, but with a base line of 500 m. The base line was increased to 5000 m around 1964. We will discuss some aspects of the operation of this system.

Requirement of High Gain with a Power Output at the Frequency f.
The converter at the outlying point furnishes a nominal power of 2 W at the frequency f_r = $f_b - f$ with an input to the mixer of the order of 1 W at f = 86 MHz. For efficient use of the nominal converter power, it is evidently necessary to provide noise amplification at the high frequency of f = 86 MHz in a band Δf = 0.5 MHz with an output power of 1 W and a gain near 10^7. Taking into account the fact that a radio-astronomy antenna is connected to the amplifier input, one can clearly see the technical complexity of eliminating positive feedback. This problem was overcome in the following manner. The necessary amplification was provided by three amplifiers having independent power supplies; one of these amplifiers was placed near the antenna, about 100 m from the converter.

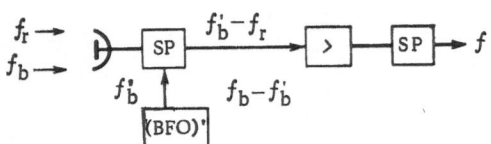

Fig. 5. Circuit with double frequency
conversion.

With an amplifier having a gain of $k \approx 10^7$ at the outlying point, a "forward-amplification receiver" could be set up for controlled reception of the sources at the radio-astronomy antenna of the outlying point.

Achieving Channeling of the Beat-Frequency Power at the Converter Output. (Outlying Point). The "Strela" transmitting unit (used as amplifier and converter) has no provision for transmission at frequencies other than 86 MHz. However, the band width of the transmitter output circuit is about 20 MHz, which is sufficiently wide that a proper tuning of this circuit will allow passage (although with some attenuation) of both f_r and f_b; here $f_b - f_r = 86$ MHz.

Achieving Power Channeling from the Converter to the Relay Antenna at the Outlying Point. The relay antennas must be placed a few meters above ground in order to reduce losses during transmission. The large size of the transmitting apparatus and the primary use conditions were such that the apparatus could not be placed in the immediate vicinity of the antenna. Connecting the antenna with the transmitter leads to a large power loss in the connecting cable. The actual power loss at the beat frequency of 2000 MHz in the cables used was about 0.11 dB/m in the RKK 5/18 and about 0.41 dB/m in the RK-3 cable.

At the 30-m Aladino tower, the cable length was 40 m of RKK and 9 m of RK-3. The losses were 4.4 and 3.7 dB — about 8.1 dB in all.

The loss could be reduced by replacing the cable with periscopic antennas, which are widely used in radio communications.

The $f_r \rightarrow f$ Conversion at Low Power Levels of the Relayed Signal and of the Beat Oscillator at the Central Point. Normal operation of the crystal mixer is usually set by the beat-frequency power, $P_b \approx 500 \, \mu$W; the transfer coefficient is about 0.1. In our case, the beat oscillator was at the outlying point and, as measurements showed, the power P_b (and also P_r) at the input of the central-point converter was of the order of a fraction of a microwatt. Calculations and experiments carried out to determine the power at the output of the crystal mixer showed that the transfer coefficient decreased to a magnitude of the order of 10^{-4} in our case.

There are two possibilities for providing normal operation of the crystal mixer.

1) Use of a centimeter-range amplifier at the converter input. This is a trivial possibility, and can be evidently realized only with some difficulty.

2) Use of a circuit* with double conversion and use of a local beat oscillator. Figure 5 shows the circuit.

It is apparently convenient to set $f_r < f_b' < f_b$. Then we have $f_b' - f_r$ and $f_b - f_b'$ smaller than $f_b - f_r = f = 86$ MHz.

On the whole, this circuit permitted the study of several features of the operation of an experimental model of a radio-relay interferometer and the preparation for a new system free of the former's shortcomings. The paper by G. I. Dobysh, included in this collection, is devoted to a description of this system.

---

*Proposed by V. V. Vitkevich.

4. The Radio-Relay Interferometer

as a Broadband Interferometer

The width of an interferometer lobe is governed by the ratio $\lambda / D \cos \varphi$. The resolution of the lobe envelope usually coincides with the resolution with the larger of the antennas.

When the frequency band of the radiation used is broadened, however, a "smearing" of the lobes far from the "zero" lobe occurs; the broader the band Δf, the greater the smearing. This is, in a certain sense, equivalent to a contraction of the lobe envelope. In this manner we obtain a new possibility for increasing the resolution, as was discussed first in [6].

Let us evaluate the equivalence in the ratio of the resolution of the lobe envelope of a broadband interferometer with a long base line and a standard antenna of size D_E. We restrict ourselves to the case of a Gaussian band of the intermediate-frequency amplifier (the discussion would be similar in the case of a rectangular band).

The modulation of the n-th lobe is of the form $M_n = e^{-\pi n \Delta f/f}$, hence the number of lobes exceeding "half power" ($M_n > 0.5$) is

$$2n = 0.44 \, f/\Delta f.$$

Accordingly, the width of the envelope at half power is

$$2n \frac{\lambda}{D \cos \varphi} = 0.44 \frac{f}{\Delta f} \frac{\lambda}{D \cos \varphi}.$$

On the other hand, the width of the directionality pattern of a standard antenna of size D_E at half power is λ / D_E. The condition for equivalent resolution is

$$0.44 \frac{f}{\Delta f} \frac{\lambda}{D \cos \varphi} = \lambda/D_E,$$

from which we find

$$D_E \approx 2D \cos \varphi \frac{\Delta f}{f}.$$

With, for example, a base line of D = 50 km, $\Delta f/f \approx 0.05$ and $\varphi \approx 0$, the envelope resolution of the interferometer is equivalent to the resolution of an antenna of size D_E = 5 km. The number of lobes in the width of the envelope at half power is $2n \approx 9$. If we are restricted to the order of ten "working lobes," then we have $\Delta f/f \approx \frac{1}{20}$. The envelope resolution of the interferometer can be increased (by the ratio $D_E \approx 0.1 \, D \cos \varphi$) by an increase in the base line D.

In the study of many discrete radio sources, the spatial selectivity determines the limiting number which can be studied with a given telescope, because of the confusion effect. For this reason, it is important to evaluate the angular separation at which two sources can be detected without mutual interference.

We assume the source under study S_n is recorded in the n-th lobe with a modulation

$$M_n = e^{-\pi n \Delta f/f}.$$

Source S_I (the interference) is of intensity β times that of the source under study. At the instant when S_n is recorded in the n-th lobe, the source S_I is recorded in the (n + n')-th lobe with an amplitude

$$\beta M_{n+n'} = \beta e^{-\pi(n+n')\Delta f/f}.$$

To solve a particular problem, we must know the degree α to which the desired signal exceeds the interfering one. We have

$$e^{-\pi n'\Delta f/f} = \alpha\beta$$

or

$$n'\frac{\Delta f}{f} = \frac{\ln(\alpha\beta)}{\pi} \, .$$

Using these expressions, we find

$$n' = n\frac{\ln(\alpha\beta)}{\ln(1/M_n)} \, .$$

This expression relates the quantities of interest to us. For example, if we have $2n = 10$, $M_n = 0.6$, $\alpha = 10$, $\beta = 1$, then we find n' = 22 with $\Delta f/f = {}^1\!/_{70}$.

This possible increase in the resolution of radio-relay interferometers seems specially important for use in overcoming confusion effects in the meter-wavelength range.

LITERATURE CITED

1. V. V. Vitkevich, in: Wideband Cruciform Radio Telescope Research (D. V. Skobel'tsyn, ed.), Consultants Bureau, New York (1969), p. 71.
2. V. V. Vitkevich, Astron. Zh., 29:454 (1952).
3. V. V. Vitkevich, Proceedings of the Fifth Conference on Cosmogony [in Russian], Izd. Akad. Nauk SSSR (1956).
4. V. V. Vitkevich, Library Report No. RF111 of the Radio-Astronomy Station, P. N. Lebedev Physics Institute, Academy of Sciences of the USSR.
5. R. Brown, R. Jennison, and M. Das Gupta, Nature, Vol. 170, No. 1061 (1952).
6. V. V. Vitkevich, Dokl. Akad. Nauk SSSR, 91:1301 (1953).

BASIC PARAMETERS OF THE EAST—WEST ANTENNA FEED OF THE LEBEDEV INSTITUTE WIDEBAND CRUCIFORM RADIO TELESCOPE

Yu. P. Ilyasov

1. Layout and Installation of Feed

The east—west antenna of the FIAN (Lebedev Institute) cruciform radio telescope comprises a parabolic cylinder, of dimensions 1008 m on the generatrix and 40 m on the tension chord, with a wideband feed installed along the focal line of the radio telescope. According to design considerations, discussed in detail in another article [1], the antenna is divided into 36 sections, each 28 m long. A section of the antenna feed, consisting of eight shunt dipoles with a wideband balancer, was installed within one such section of the antenna. The parameters of the antenna section have been reported [2]. That article also showed that the antenna feed section allows the antenna to operate simultaneously over a broad frequency range, stretching from 30 to 120 MHz.

RKK-5/18 grade down-lead cables, each 50 m long, were laid along the dipole masts from each feed section. The down-lead cables were all of the same length, to within ±5 mm. The down-lead cables from two adjoining sections were led to the summing wideband transformers, which are in the form of tapered coaxial lines 4 m in length. These transformers are installed directly at the ground level. From the summing transformers, RKM-5/18 grade cables each 465 m in length lead to the center of the antenna. There are 18 such cables in all, nine on each side of the center of the antenna. They are laid underground at a depth of about 1 m in special asbestos-cement tubes, in order to minimize any chance of their electrical length changing in response to temperature fluctuations. Excess cable lengths from the central sections of the antenna feed are wound on drums placed at the center of the antenna, in a special basement switch house. Lines of variable length installed at the terminations of the 18 cables in this switch house are employed in order to equalize the electrical lengths of these cables. All sections of the feed are combined later on as in the diagram in Fig. 1, using summing wideband transformers with resistance transforms from 25 to 75 Ω. This is achieved by the use of tapered coaxial lines 4 m in length. With this antenna switching arrangement, uniform in-phase excitation of all sections of the feed is achieved with excellent impedance matching over a wide range of frequencies.

We clearly realize from Fig. 1 that the switching arrangement resorted to here makes it possible to operate with either compensation-type radiometers or correlation-type radiometers. In addition, this arrangement also makes it possible to form interferometers with any desired base line within the limits of the entire antenna array. It has been demonstrated [3] that deviations in the directional behavior of the radiation pattern in response to random phase errors

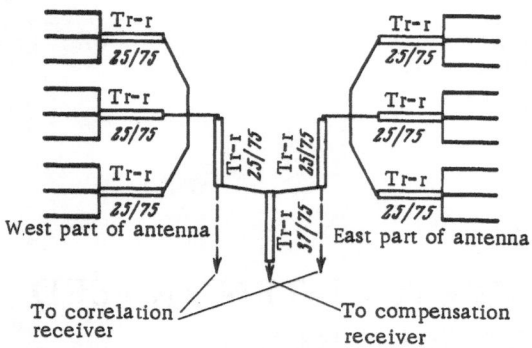

Fig. 1. Switching arrangement for feed
sections.

in the feeder lines of the antenna will be smaller in the case of this switching arrangement than in the case of one with a smaller number of independent high-frequency cables leading to the center of the antenna for summation.

2. Antenna Feed Adjustments

Adjustment of most of the feed components is carried out before they are mounted on the antenna. The shunt dipoles are assembled on a special forming board. After assembly, dimensional tolerances are checked. Deviations in dimensions were not greater than ±1 cm. Prior to assembly of sections from the eight dipoles, all segments of the feed line leading to that point were first adjusted against a single standard. The assembled section was then given a final tuning and adjustment, with matching and phasing measured over a broad frequency range. Phasing accuracy on absorbing loads included in place of the dipoles during the tuning process [2] was to within ±3°. After a final check without any dismantling, the section of the feed was then mounted in the reflector. Wideband balancers and summing transformers were checked over the 30 to 120 MHz frequency range for impedance matching and amplitude — phase symmetry at the output. All the RK-3 cables joining the various elements of the feed lines were tuned against unified standards to within ±3 mm, by the usual procedure using the LI-3 slotted measuring line.

The main feed lines, laid underground, were checked and tuned with special care. Tuning was initially by a pulsed technique. Pulses 0.1 μsec in duration were supplied simultaneously to a "standard" cable and to the cable being measured. The cable being measured was equalized to within ±60 cm with respect to the standard cable, on the basis of the pulses reflected from the terminations of the cables. A DÉSO-1 high-speed dual-beam oscillograph was employed in these measurements. All inhomogeneities in the cable measured which yielded reflections greater than 0.1 were first removed. Their positions could be determined with ease from time marks on the DÉSO-1. As a rule, most of the inhomogeneities appeared at points where different structural lengths of the cables joined. Accurate pretuning of the cables to within ±60 cm was determined by the duration of the leading edge of a main pulse, the maximum sweep rate of the DÉSO-1, and the scale of the DÉSO-1 time markers. In our opinion, the attainable accuracy in the readings was in fact achieved, since the leading-edge duration of the reflected pulse depends on the length of the cable, because of the frequency variation of attenuation in the cable. Calculations showed that a rectangular main pulse reflected back from the cable termination in RKM-5/18 cable about 500 m long would be returned undistorted with a wave front duration on the order of 0.075 μsec.

The pulsed technique also proved convenient for rough checks on identical attenuation in the cables. At slow sweep rates, multiple reflections from the two terminations of the cable could be observed, their envelope yielding an exponent. Absence of contacts in distinct sections of the feed could be detected by repeated reflections in the down-lead cable during the tests, and thereby eliminated. In that sense, the pulsed technique can be proposed as a convenient system for inspecting feed lines of the type described.

Definitive phase correction of the underground cables involved the use of a modulation method described in detail in another article [4]. This method of measurement makes it possible to equalize the electrical lengths of the cables unambiguously within the range of a half-wavelength difference in the lengths of the feed lines. The principle advantage of this modula-

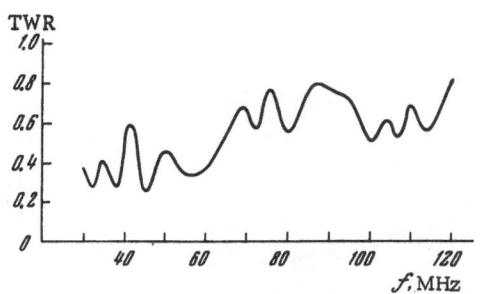

Fig. 2. Matching of one of the feed sections. TWR = $F(f)$.

tion technique lies in the fact that it can be used for phase adjustments of long lossy transmission lines to a high order of precision. In our case, the equivalent attenuation due to the cable and the measurement circuit was about 40 dB. For a load with a modulated reflection coefficient, we used a T-junction with a matched load inserted in one arm and a 1A501 semiconductor diode in the other arm. The modulation voltage was placed across the semiconductor diode, from a portable transistor audio oscillator. Prior to the measurements, the modulated loads of the cable to be measured and the "standard" cable were checked and the initial difference in their electrical lengths was determined. The method provided accuracy to within ±0.5 cm in measurements at a single frequency.

It was shown that the measurements can also be carried out with the same modulated loads at wavelengths in the decimeter range. But the real accuracy attainable in tuning is limited not by the measurement technique, but by small inhomogeneities in the cable. These inhomogeneities, characterized by a reflection coefficient not greater than 0.1, yielded deviations in electrical length in the ±5° range at different frequencies (about 5 cm at 3.6 m wavelength), because they are located at different positions along the feed line. In consequence, the measurements had to be performed at three nonmultiple frequencies, and tuning adjustments were made with respect to the mid-difference in lengths.

The modulation method proved to be suitable for measuring small inhomogeneities in lossy feed lines by the method of the displaced shorted piston. A well-matched feed line of variable length was inserted between the modulated load and the cable in order to facilitate these measurements. These two techniques were relied upon for tuning all the underground cables to within ±5 cm electrical length (relative error $\pm 10^{-4}$) with inhomogeneities in the line exhibiting a reflection coefficient not greater than 0.1. Measurements of attenuation using a radiometer and a portable noise generator showed the deviations in cable attenuation to be not more than 20% off tabulated values.

Matching of all sections of the feed was measured over a wide range of frequencies in the cross section at the first summing transformer. It can be safely stated, to a degree of accuracy consonant with the errors in the measurements, that the results of the measurements are in agreement. One of the curves plotted (results of measurement of the matching of section No. 7, as numbered from the west end of the antenna) is shown in Fig. 2.

Efficiencies were computed for the feed section, and for the entire feed as a unit, on the basis of results of measurements of the matched terminations, and on the basis of attenuation data for the cables and the measured crosstalk attenuation in the feed line devices over the range of frequencies of interest here. The results of these calculations are plotted in Fig. 3. The dashed curves are plotted here for the case of completely matched dipoles and feed lines over the entire frequency range. Efficiencies measured at several frequencies in terms of galactic radio emission and averaged for the several sections, are plotted in Fig. 3a. The agreement between the experimental values and the theoretically predicted values is quite satisfactory.

The radiation patterns in the E-plane of some of the antenna sections were measured at different frequencies in the operating range. The measurements were based on radio emission from Cassiopeia A. Results of the measurement showed that the shape of the radiation pattern

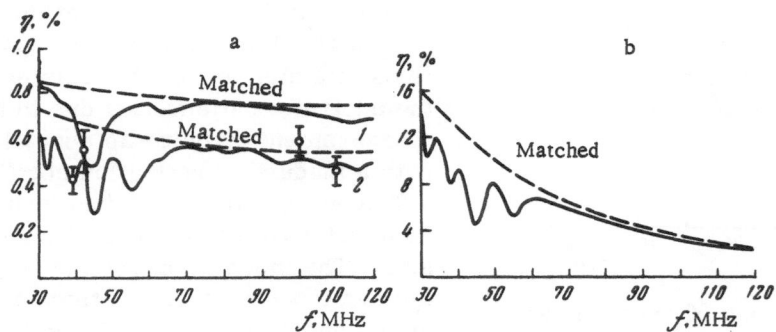

Fig. 3. Efficiency of antenna feed. a) Efficiency of feed section
(1) in cross section with wideband balance, and (2) in cross section with transformer; b) efficiency of feed line.

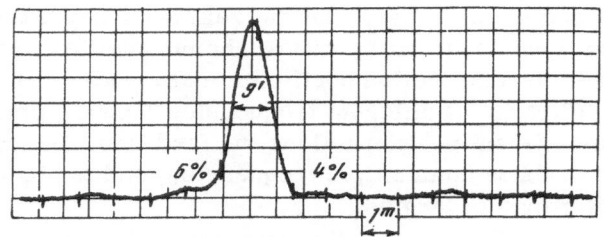

Fig. 4. Radiation pattern of the east — west antenna in the E-plane
at 110 MHz frequency.

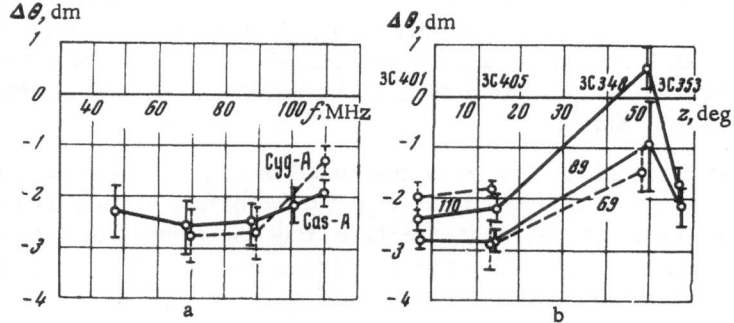

Fig. 5. Results of measurement of the position of the radiation pattern. a) As a function of frequency; b) as a function of zenith distance (computed inclination of east — west axis to horizon 30°).

was close to the calculated shape, and that the effective area corresponded to that expected at a
reflector with a coefficient of usable surface of the order of 0.5.

       After all the underground cables had been tuned, the parts of the antenna were summed
and the radiation patterns of those parts were checked in stages of summation as indicated in
Fig. 1. The radiation patterns were measured against several powerful sources of radio emission. Figure 4 shows a radiation pattern for the entire antenna, in the E-plane, obtained at 110
MHz frequency using the source Cassiopeia A. The half-power width of the radiation pattern was
9'.0 ± 0'.3 (calculated with 9'.5), and the side-lobe was not higher than 6% (calculated level 4%).

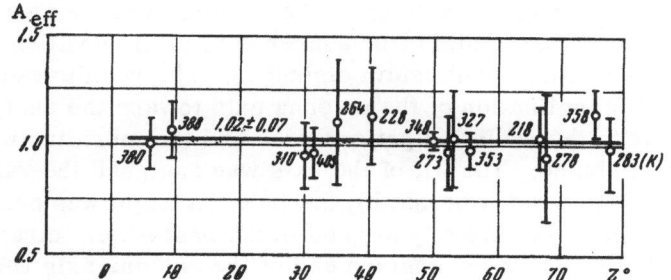

Fig. 6. Dependence of effective antenna area on zenith distance. Numerals at data points indicate number assigned to source in 3C catalog.

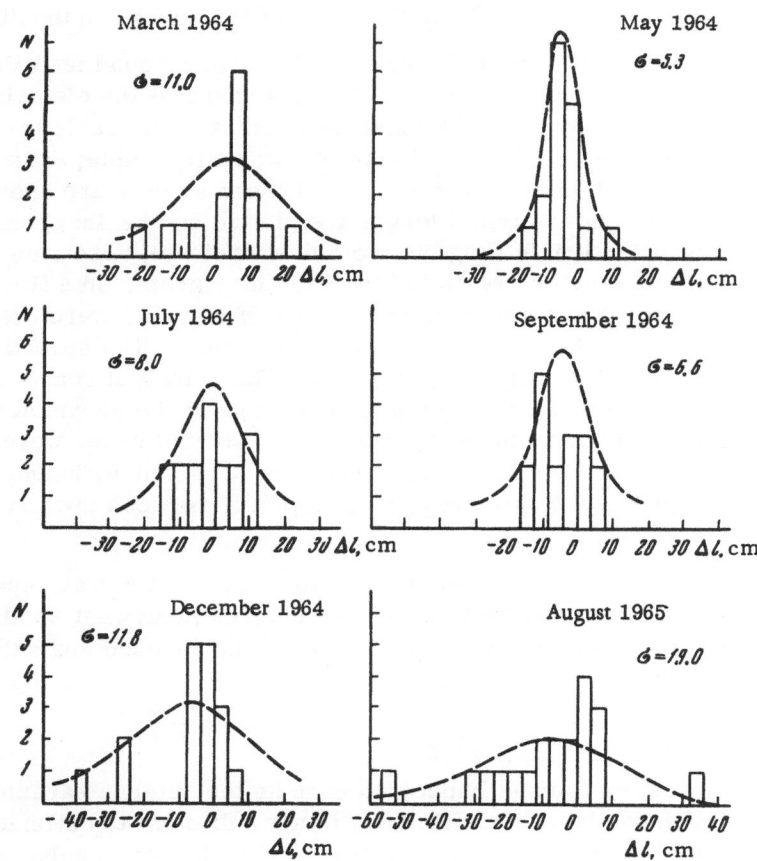

Fig. 7. Histograms of the spread in electrical lengths of underground cables.

The E-plane measurements of the radiation patterns at other frequencies within the operating range showed the parameter patterns to be close to calculated values at those frequencies as well.

The location of the peak on the radiation pattern at different frequencies was measured using the sources Cassiopeia A and Cygnus A, in order to obtain information on the phase behavior of the feed over a wide frequency range. The results of these measurements are plotted in Fig. 5a. Figure 5b gives the results of measurements of the position of the radiation pattern at different frequencies and different zenith distances. The results were processed with attention given to the existing inclination of the antenna axis toward the east. As a result of this antenna tilt, the angle between the radiation pattern peak and the meridian will vary according to the cosine of the zenith distance. The tilt of the axis was assigned the value 30' in the calculations. As measurements showed (Fig. 5a, b), the true tilt angle was near 28'. As is clear from Fig. 5b, the antenna axis is not strictly aligned in the east — west direction, but deviates from that direction in such a way that the eastern end of the antenna axis veers northward by an angle of about 1'. The subsequent measurements based on a large number of sources [5] yielded the following values: tilt of axis to horizon, 29'.58 ± 0'.39; rotation of axis, 1'.86 ± 0'.43, i.e., the antenna diagram on the horizon deviates from the meridians eastward by that angle.

Measurements of the effective antenna area based on intense sources at 3.5 m wavelength showed that no significant changes in the effective area were detected, to within 7% error, in the case of zenith distances from 10° to 78°. Figure 6 shows the results of these measurements. The numerals at the data points indicate the numeration of the source in the 3C catalog.

Regular verifications of the electrical lengths of the underground feed lines were carried out while the east — west antenna was in service. It was found that the electrical lengths of the transmission lines varied with respect to the "standard" cable. The cable proceeding to the central section of the antenna was assigned the role of "standard" cable, since it lay almost entirely within the basement switch house. The results of these studies are shown in a series of histograms in Fig. 7. Changes in electrical length are plotted as abscissa, and the number of cables not exceeding a specified length range is plotted as ordinate. The rms deviations (σ) of each distribution of electrical lengths are stated for each histogram. In a first approximation, it is safe to say that no seasonal variations or periodicity of any kind were observed. Cable lengths were readjusted after each check on the spread of values. The spread in values of cable lengths increased with the time between readjustments. The principal reason for the change in the electrical lengths of the cables is, in our view, the change in the parameters of small inhomogeneities, at locations of high-frequency connectors. Several cases were reported where, for inhomogeneities revealed in the course of checks by a pulsed technique, deviations in electrical length on the order of 20-30 cm were drastically reduced upon removal of the inhomogeneities.

A change in antenna parameters due to changes in cable lengths was reported. The sidelobe level increased to 8-10%, while the effective area became somewhat smaller. As a result, it was recommended that the electrical length of the cables be checked and adjusted not less than once every three months.

3. Sensitivity of the East—West Antenna

The effective antenna area is usually decided upon as the antenna parameter. However, a knowledge of the effective area and of the efficiency is not sufficient to estimate the sensitivity of the antenna. The noise temperature of the antenna must be known to solve problems in radio astronomy [6].

In the meter-wavelength range, the antenna noise temperature is determined primarily by the brightness temperature of the galactic background radio emission, which declines rapidly as the frequency rises [7]. Since the brightness temperature differs over the celestial sphere, we assume peak temperatures for the minimal sensitivity (in the direction pointing toward the center of the galaxy) and minimum values for the maximum sensitivity (in the direction pointing toward the galactic pole), in the calculations. Moreover, it has been shown [8, 9] that faint

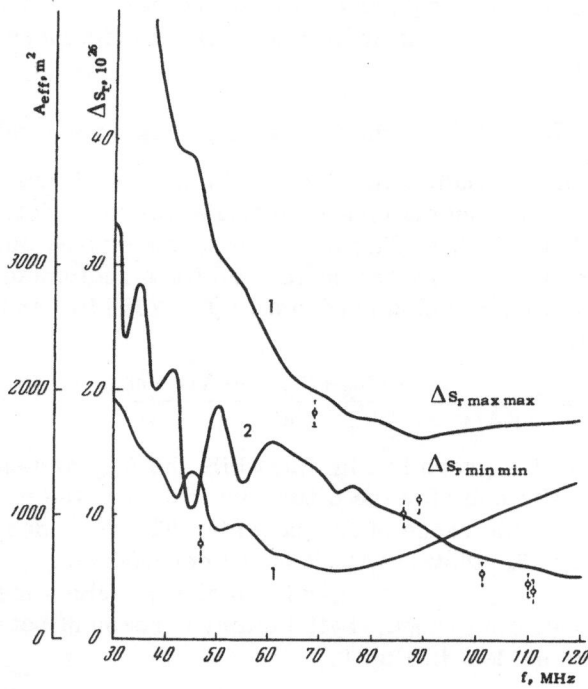

Fig. 8. Dependence of (1) the limiting sensitivity of the east—west antenna and of (2) the effective antenna area A_{eff}, on the frequency.

TABLE 1

f, MHz	T_s, °K		T_{bg}, °K	
	min	max	min	max
30	40 500	82 500	10 000	120 000
40	22 500	54 000	5 000	61 000
50	12 000	40 500	2 900	37 000
60	6 000	30 000	1 900	24 500
70	1 500	24 000	1 300	17 500
80	0	19 500	960	13 000
90	0	15 000	720	10 000
100	0	12 800	570	7 800
110	0	9 500	420	6 000
120	0	7 800	360	5 000

discrete sources appearing simultaneously within the field of reception yield fluctuations in total flux due to fluctuations in the number of such sources. This obviously means fluctuations in the readings of the output recording instrument, and will also determine the limiting sensitivity of the radio telescope. Note that the characteristic period of these fluctuations. is determined by the time it takes sources at a given declination to travel across the antenna radiation-reception pattern. The equivalent noise temperature has been shown [9] to be a convenient characteristic of these fluctuations in estimating the sensitivity of a radio telescope. Equivalent noise temperatures were calculated for two alternative possible laws of dependence of the number of discrete sources on source intensity, and will be denoted below as T_s^{max} and T_s^{min}.

Values of these temperatures for the frequency range of the east — west antenna are tabulated in Table 1.* The table lists values of the maximum and minimum brightness temperatures of galactic background emission [7].

In determining $T_s = \frac{\Delta S \, A_{eff} \sqrt{\Delta f \tau}}{2k}$ the following values were used: effective antenna area $A_{eff} = 0.6 \, A_{geom} = 24{,}000 \text{ m}^2$; radiometer bandwidth $\Delta f = 300$ kHz; and time constant $\tau = 10$ sec. The rms values of the fluctuations in average flux were determined from data reported in [9]. Calculations of the limiting flux from a discrete source picked up by a radio telescope and at a signal-to-noise ratio of 5 were performed for a radiometer with the parameters $T_r = 500°$K (noise factor 2.6), transmission bandwidth $\Delta f = 300$ kHz, and time constant $\tau = 10$ sec, using the familiar formula

$$\Delta S_r = 5 \frac{2k}{\sqrt{\Delta f \tau}} \frac{T_r + (T_{bg} + T_s)\eta + T_0(1 - \eta)}{A_{eff}} \ .$$

The efficiencies η are read off the graph in Fig. 3b, while the A_{eff} values at the receiver input were computed by using this efficiency for the utilization ratios of the surface, which fall within the interval 0.6 to 0.53 for a wide range of frequencies. The dependences of $\Delta S_{r\,min\,min}$ and $\Delta S_{r\,max\,max}$ are plotted in Fig. 8. Plotted with it is the dependence A_{eff} over a broad range of frequencies. The points on the graph denote experimental A_{eff} values measured at distinct frequencies within the antenna operating range. Satisfactory agreement between the experimental values and predicted values is evident in Fig. 8.

It is also clear in Fig. 8 that antenna sensitivity is determined primarily by background radio emission and by the saturation effect, in the 30–70 MHz frequency range. Even though the effective antenna area at the receiver input contracts with rise in frequency, because of losses in the feed lines, the sensitivity of the radio telescope comes out improved. In the 80–120 MHz frequency range, antenna sensitivity with frequency rise is even more affected by the parameters of the feed lines. Consequently, the use of outboard amplifiers [9] means a gain in the 80–120 MHz range, but virtually no improvement in sensitivity at all in the 30–70 MHz range. And in that case determination of the sensitivity of the radio telescope appears, in our view, to demand treatment of all factors pertinent to the operating range of frequencies. Estimates of antenna performance based on the effective antenna area alone would be inadequate. Observations carried out at 7.5-m wavelength have shown that the sensitivity of the radio telescope is in fact determined by background radio emission and by the saturation effect.

By designing a wideband feed with a 4 : 1 frequency overlap, we were able to make full utilization of the antenna's capabilities. There are a lot of problems which are difficult to solve with a narrowband instrument in the case of this type of radio telescope. To be specific, measurements of the spectral indices of many sources [5] can be carried out most simply on a single wideband instrument. For another thing, high stability to noise exhibited by wideband antennas makes it possible to obtain more information and to more fully utilize the expensive instrument.

In conclusion, the author welcomes this opportunity to express his gratitude to the leader of the project, V. V. Vitkevich, for his persistent attention to the progress of the work, A. D. Kuz'min for highly appreciated comments at the start of the work, and V. T. Solodkov for invaluable assistance in performing the experiment.

---

*These T_s values were obtained by R. D. Dagkesamanskii.

LITERATURE CITED

1. V. V. Vitkevich and P. D. Kalachev, in: Radio Telescopes (D. V. Skobel'tsyn, ed.), Consultants Bureau, New York (1966), p. 1.
2. Yu. P. Ilyasov and A. D. Kuz'min, in: Radio Telescopes (D. V. Skobel'tsyn, ed.), Consultants Bureau, New York (1966), p. 7.
3. S. N. Ivanov, Yu. P. Ilyasov, and G. N. Khramov, in: Radio Telescopes (D. V. Skobel'tsyn, ed.), Consultants Bureau, New York (1966), p. 13.
4. G. Swarup and K. S. Yang, IRE Trans. on Antennas and Propagation, AP-9, No. 1 (1961).
5. V. S. Artyukh, V. V. Vitkevich, and R. D. Dagkesamanskii, Astron. Zh., Vol. 44, No. 5 (1967).
6. L. D. Bakhrakh and K. I. Mogil'nikova, Izv. Vuz. Radiofiz., 7:585 (1964).
7. H. K. Sutcliffe, Proc. IRE, Vol. 51, No. 10 (October, 1963).
8. S. von Hoerner, NRAO, Vol. 1, No. 2 (1961).
9. R. D. Dagkesamanskii, S. N. Ivanov, and Yu. P. Ilyasov, in: Wideband Cruciform Radio Telescope Research (D. V. Skobel'tsyn, ed.), Consultants Bureau, New York (1969), p. 25.

APPARATUS AND METHODS USED
IN RADIO-ASTRONOMICAL MEASUREMENTS
OF THE VELOCITY OF THE SOLAR WIND

I. A. Alekseev, V. V. Vitkevich,
V. I. Vlasov, Yu. P. Ilyasov,
S. M. Kutuzov, and M. M. Tyaptin

1. Investigation of Flickering of Radio Sources and the Necessity for Synchronized Observations at Three Points

Starting in 1951, when studies of the medium around the sun by the "transillumination" method first began [1], a very large number of investigations on the scatter of radio waves in the circumsolar plasma have been carried out. The chief parameter which has been measured and investigated is the angle of scatter of the radio waves during their spread through nonhomogeneities. From data relating to this scatter, the characteristics of nonhomogeneities of the supercorona of the sun have been obtained [2-8].

Under certain conditions flickerings of radio waves are observed, i.e., an alternating intensity is recorded. Changes in the intensity (or, more exactly, the density) of the energy flux at the observation point are accompanied, of course, by changes in the angular spectrum, i.e., by scatter. However, we shall concentrate attention on the change in intensity.

Observations of flickering of radio waves crossing large-scale nonhomogeneities were mentioned originally in 1955 [2]. Here, however, we shall discuss only the flickering caused by small-scale nonhomogeneities, filling the whole space about the sun, and always existing. Let us consider the chief mechanism of origin of the flickerings.

In Fig. 1 the arrows IW show the directions of movement of corresponding parts of the incident wave front from an infinitely distant point source. The wave front is a plane and all the arrows are parallel to one another. If we mentally draw the trajectory of the rays through a medium filled with nonhomogeneities, for each trajectory, although the geometrical paths are equal, the electrical paths will generally speaking be different, on account of the presence of a varied number of nonhomogeneities with a refractive index different from unity. When it emerges from the layer containing nonhomogeneities, the front of the disturbed wave will therefore have irregular directions in certain areas. In our case, all directions will differ only a little from the principal direction of spread of the wave along a straight line in the absence of nonhomogeneities.

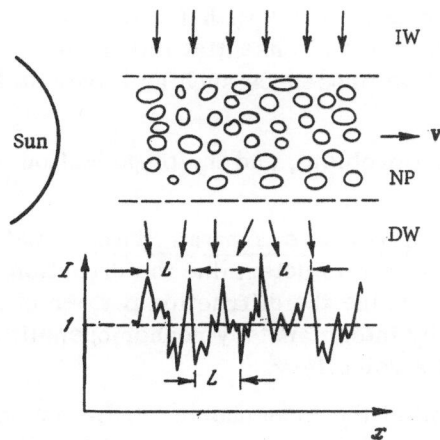

Fig. 1. Scheme of formation of diffraction pattern and of flickerings. IW) Flat incident wave; NP) nonhomogeneous plasma-disturbing layer; DW) disturbed wave; **v**) direction of velocity of movement of nonhomogeneities. The diffraction pattern in the plane of the drawing is shown in the bottom graph.

As a result of interference between individual (conventional) rays the irregular wave front, as it spreads, will create an irregular, not homogeneous, field in the plane perpendicular to the direction of spread of the wave. If we intersect this field in the plane of the drawing, in a direction perpendicular to the spread of the wave, we obtain an irregular distribution of intensity I along the x coordinate, as shown, for example, in the drawing.

If, however, the relative position of the source of each of the nonhomogeneities and of the observer remained unchanged, a certain constant intensity I(x) would be recorded at the point of reception, and this could be greater or less than the mean. However, the situation is otherwise. First, the source moves relative to the nonhomogeneities and to the earth; second, its position changes relative to the position of individual nonhomogeneities; third, all nonhomogeneities move, together with the circumsolar plasma, outward from the sun; and, fourth, the earth revolves and rotates relative to the nonhomogeneities. The result of all these factors is that the diffraction pattern of the field undergoes changes in time and moves as a single entity relative to the observer, thus creating flickering of the radio sources, i.e., a change of intensity in time at a given point.

Of all the above factors causing flickering, the most important is the presence of radial movement, common to all the nonhomogeneities, outward from the sun with velocity **v**. Determination of a reliable value of **v** would enable consistent values to be obtained for the linear dimensions of the diffraction pattern, and also, consequently, for the dimensions of the nonhomogeneities; this parameter, in turn, would enable the electron concentrations of the nonhomogeneities to be determined.

Besides determination of the value of the dimension a, another extremely important result may be mentioned, which can be obtained by investigating the relationship between **v** and the distance from the sun, if it can be shown that the nonhomogeneities are "frozen" into the plasma. This brings us to the nature of solar wind.

At present there are only fragmentary experimental data concerning movement of the plasma close to the earth's orbit (in the interval as far as Venus) obtained from rocket measurements, and also data for the wind velocity near the sun itself, obtained by the radar method. This information is not enough to provide the answer to one of the main problems in the physics of interplanetary plasma: the nature of the solar wind. According to Parker's theory, the velocity of the wind increases substantially with an increase in distance to about 50 $R_\odot$, and then remains practically unchanged.

If, however, the movement is caused by discharges (the sum of the "flares"), the velocity will decrease with an increase in the distance. Probably in either case near the earth's orbit the change in velocity will be small, and it will hardly be possible to observe the character of the change in velocity from rocket measurements made between the earth and Venus. On the other hand, analysis of the velocity near to the sun would not only confirm the theory, but would also enable the acceleration to be determined near to the sun, i.e., it would yield new parameters.

The question whether the solar wind is in reality produced in accordance with Parker's theory, or whether the nature of the forces creating it are different, is one of essential importance. As yet no definite answer can be given to it. Very probably, active processes such as flares on the sun play the chief role in the formation of solar wind.

The study of solar wind is thus itself a very interesting problem, and the organization of appropriate experiments has been undertaken.

Organization of Experiments and Demands on the Apparatus. Let us discuss some aspects of the organization of observations in order to determine the direction and velocity of movement of the nonhomogeneities. Let us imagine the diffraction pattern of the field on the earth with characteristic dimension L, created by interplanetary nonhomogeneities of the plasma, moving relative to the observation point with a velocity v.

In that case, just as is done in relation to measurement of the ionospheric wind, three observation points can be arranged, forming approximately an equilateral triangle. Let the distance between them $d \ll L$, so that there is a high degree of correlation, close to unity, between the observed fluctuations of energy at the different points.

If α_{ik} represents the angle between the velocity vector and the direction of the base, the difference between the times of appearance of the same fluctuation changes of intensity at each point will be given by $t_{ik} = d_{ik}/v \cos \alpha_{ik}$. If it is taken that $\cos \alpha_{ik} \approx 1$, then for the characteristic velocity $v \approx 300$ km/sec, the distances $d_{ik} = 300$ km will correspond to a time of delay of 1 sec. The rational choice of the distances d_{ik} is an important question. If d_{ik} is too great, correlation between the fluctuations at the two points is disturbed, while low values of d_{ik} lead to a small change in the pattern with time, i.e., to a low accuracy in the measurements.

We make the initial assumption that the nonhomogeneities on which diffraction takes place move relative to the earth, and undergo little change during time T between the maxima of fluctuating intensity. We thus consider that the fluctuational picture which we observe is due mainly, not to changes in the nonhomogeneities themselves, but to the general movement of the diffraction pattern relative to the earth. If τ_n represents the life span of one nonhomogeneity, it can be accepted that $\tau_n \gg T$. We thus consider that τ_n is of the order of 10 sec or more.

If the radial velocity relative to the sun is the same for all nonhomogeneities, then because the diffraction pattern is formed as the result of the action of all the nonhomogeneities, it will be deformed by the relative displacement of the nonhomogeneities lying in the line of sight. If β is the angle within which the nonhomogeneities make their main contribution to flickering, the tangential velocity along the line of sight will vary within the limits $v_t = v(1 \pm \cos \beta)$. If, for example, $\beta = 30°$, then $\Delta v = \pm 14\%$. If the nonhomogeneities move through the characteristic distance L, the relative displacement will be $\pm 0.14L$. If, therefore, the distance between the points of observation is of the order of L, correlation will not be disturbed on this account. If $T \approx 3$ sec and $v \approx 300$ km/sec, then L = 900 km and the points of observation could be spaced so far apart. However, this is a very great distance, unsuitable for the first observations and, furthermore, under these circumstances velocities of lower magnitude could not be measured. In addition, the solar wind along the line of sight is hardly likely to have the same magnitude over a wide range of values of β.

Estimation of the shortest distance of dispersal of the points is determined by the wind velocity and the minimum error with which the delay time between two points of observation can be measured. The probable error Δt is determined by the relationship

$$\Delta t = \sqrt{\Delta t_1^2 + \Delta t_2^2 + \Delta t_3^2}.$$

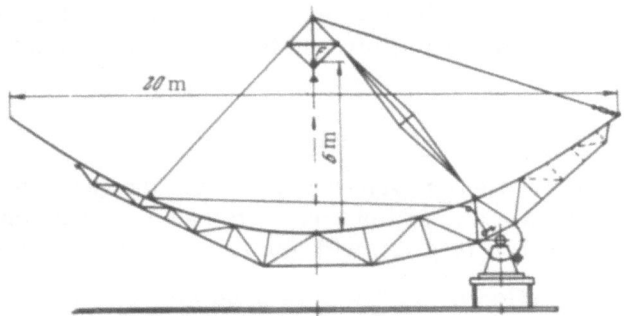

Fig. 2. Diagram of cross-section and chief measurements of antenna of extended radio telescope.

Fig. 3. General view of extended antenna.

Here Δt_1 represents the error due to differences between the apparatus at different points. If accurate calibration is carried out at the beginning of the work, it may change in the course of time. If no special measures are taken to stabilize the instruments, for the estimates it can be taken that $\Delta t_1 \approx 0.1$ sec. The term Δt_2 represents the errors in synchronization of the two points. With careful adjustment of the system this can be reduced to a minimum, and it can evidently be disregarded. The term Δt_3 is the error caused by differences in the character of recordings of the signal at different points, arising as a result of: a) disturbance of coherence of the recordings at separate points, and b) the presence of an interfering action of apparatus noise and external sources. The importance of this last factor will be influenced, in particular, by the efficiency of the antenna, the sensitivity of the receiver, the intensity of the observed source, and the degree of its radio flickerings. On the basis of present experience with observations of this type, and assuming that the effective area of the system is about 10^3 m^2, it can be taken that $\Delta t_3 = 0.15$ sec for sources 3C-48, 3C-147, and others of similar intensity and with a flicker of about 30%. We thus obtain that $\Delta t \approx 0.15$-0.2 sec.

To take $\Delta t = 0.2$ sec means that a distance $d \approx 200$ km must be chosen. This is a suitable distance for working conditions and it enables reliable measurements to be made between limits of velocity of 400 and 50 km/sec.

After careful reconnaissance the following sites were chosen for the new observation points: the first point near the town of Kalinin and the second near the town of Pereslavl'-Zalesskii.

2. Antennas of Extension Radio Telescopes

Construction of Extension Antennas.   As the two extension antennas for meter waves for work in conjunction with a cruciform radio telescope it was decided to develop an antenna of the parabolic cylinder type (Fig. 2), with the following parameters:

1) Geometrical area of reflector, $280 \times 20 = 5600$ m^2;

2) angular aperture of the parabola forming the parabolic cylinder, 159°14';

3) focal length, 6.0 m; and

4) limiting angles of rotation of antenna about the angle of elevation from the zenith (a) to the north 10°; (b) to the south 70°.

The cross section of the antenna is shown diagrammatically in Fig. 2.

One of the chief conditions to be observed during development of the project was the necessity of building antennas of simple design, with minimal loss of time in assembly. The design of the antennas was to be such that they were as transportable as possible; the possibility that the antennas might be moved in the future to other bases had to be provided for.

The reflector of the antenna consists of 130 stretched steel wires, fixed to supports comprising nine parabolic trusses. These trusses are arranged in a line along the axis of rotation, at a distance of 35 m apart, and they form a parabolic cylinder 280 m in length (Fig. 3).

The parabolic trusses are three-dimensional and consist of three members. The cross section of the trusses is triangular and the position of the axis of rotation is asymmetric. Because of this last feature, the antenna can be placed at a low height from the ground, which not only greatly facilitated its assembly, but also enabled a large sector of rotation of the antenna to be obtained about the angle of elevation.

The rotating dipole masts to which the dipoles are fixed, the subreflector, and the feeder make it possible to mount the exciter on the ground, and then to fit it in its assembled form in the focus.

The wires of the reflector are fixed in holes in the upper struts of the trusses by means of rubber bushings, which damp most of the vibration of the wires produced by wind loads. To prevent breaking of the wires from nonalignment of the trusses during manual rotation, taut steel wires are fixed to the ends of the trusses, and these prevent excessive misalignment. To damp vibration at the points where the wires are attached to the end trusses, crosswires are interwoven with them.

The end parabolic trusses are similar in construction to the intermediate ones, but to take the loads from the tension in the wires and cables they are strengthened and are held by special stays and tightening devices.

The rotating mechanism consists of a system of cylindrical cogwheels, one pair of worm gears and chain transmission from a manual drive. The rotating mechanisms are mounted on platforms with their foundations on piles.

Taken as a whole, the design of the extended antennas is relatively simple, they are cheap to build, and they have good parameters of performance.

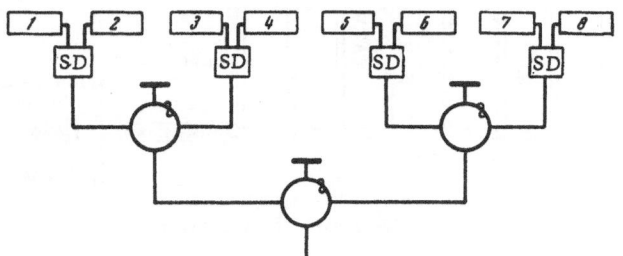

Fig. 4. Block diagram of the emitter. SD) symmetrizing device.

General Electrical Characteristics of the Antennas. As was stated
above, both antennas at the out-stations consist of parabolic cylinders measuring 280 × 20 m.
If uniformly excited, the polar diagrams of the antennas at half power have the following measurements: 42' × 10° for a wave of 3.5 m and 84' × 20° for a wave of 7 m in planes E and H, respectively.

It was mentioned previously that for simultaneous observation of a radio source at all
three points, the polar diagrams of the antennas at the out-stations must be inclined to the in-
phase position by an angle determined by the longitude of the particular station and the angle of
inclination of the observed source. For the selected sources the direction diagram of the antenna at the out-station in Kalinin Region must be inclined at an angle of 2.5° to the east, and the
direction diagram of the antenna at the out-station in Yaroslav Region by 1° to the west. This
inclination is carried out by including phase-shift controls in the corresponding branches of the
circuit of the dipole-feeder system.

The antenna exciter is composed of dipoles placed along the focal line of the reflector.
To form the diagram of the exciter in a plane H a subreflector with a width of 1.65 m and made
of wires 2 mm in diameter is used. The height of suspension of the dipoles above the subreflector is 1.05 m. The dipoles and feeder system of the exciter are mounted on different sides
of the subreflector to reduce their mutual effect.

Block Diagram of the Dipole-Feeder System. The whole antenna is di-
vided into four sections. Down-lead cables of type RKM-5/18, 50 m in length, are laid from
each section of the exciter on the dipole masts. The down-lead devices are combined in pairs
by means of wideband doublet rings [9]. The doublet rings are followed by two RKM-5/18 cables,
80 m long. If necessary they are joined by a wideband doublet ring. The block diagram of the
antenna exciter is shown in Fig. 4. The numbers denote the serial numbers of the spans. All
components of the feeder circuit were tested and tuned to a wavelength of 3.5 m. Phase adjustment was carried out so that the total error does not exceed 3-4° on wavelengths of either 3.5
m or 7 m.

The Antenna Exciter Section. The exciter section consists of two identical
spans. Ten dipoles are mounted in each span of the atenna, 35 m in length. The dipole is whole-
wave at a wavelength of 3.5 m and half-wave at a wavelength of 7 m. The wave impedance of
such a dipole is 450 Ω. The input impedance of one dipole, fixed above the subreflector, was
measured in the frequency bands from 41 to 46 MHz and from 82 to 88 MHz.

To simplify the feeder system in the span a scheme with a series power supply to the dipoles was chosen. The main twin-wire feeder with wave impedance of 490 Ω passes through the
whole span beyond the subreflector. The dipoles are connected to the twin-wire feeder by
lengths of KATV cable so that the distribution of current along them at wavelengths of 3.5 and
7.0 m is in-phase. For this purpose the points of connection of the KATV to the twin-wire feeder
were spaced 3.5 m apart, constituting one wave at 3.5 m and one half-wave at 7.0 m.

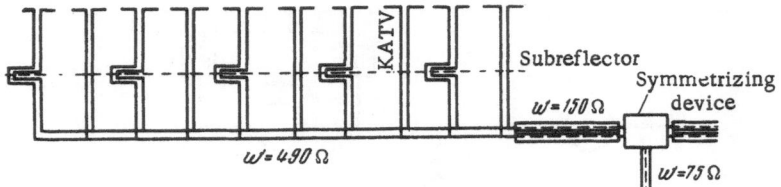

Fig. 5. Scheme of one span.

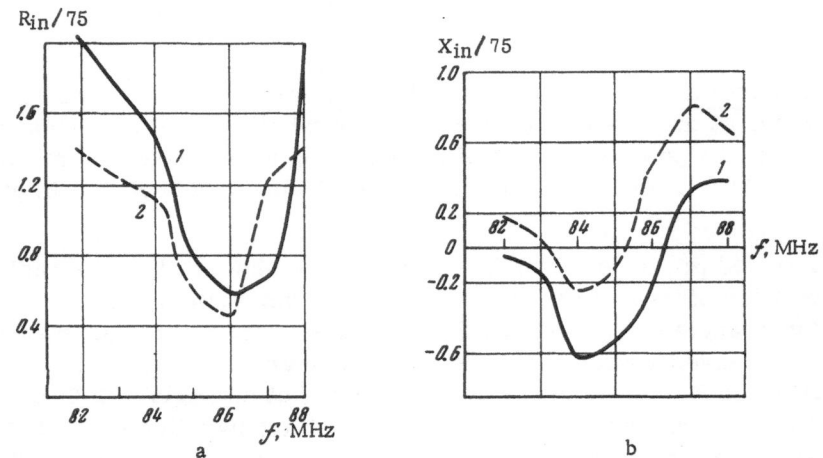

Fig. 6. Input impedance of one section in the 82-88 MHz band.
a) Active component; b) reactive component. 1) Calculation;
2) experiment.

In-phase excitation of all dipoles in the span to both waves is ensured by the appropriate choice of electrical lengths of the KATV cables. If the KATV lengths chosen were equal, at a wavelength of 7 m the alternate dipoles are excited in opposite phases. To prevent this, the lengths of alternate KATV cables are increased by 3.5 m, i.e., by a half-wave for a wavelength of 7 m. Obviously at a wavelength of 3.5 m in-phase excitation is undisturbed. These additional lengths of the KATV cables are introduced as loops fixed in the plane of the subreflector.

The principles of design of the span are illustrated in Fig. 5. It must be emphasized that this scheme of series connection of the dipoles is strictly in-phase only at wavelengths of 3.5 and 7 m. The polar diagram deviates from an in-phase state by a distance of not more than 0.1 of its width during a change of frequency of not more than 1.75%.

As the measurements showed, the dipole has a high active input impedance at a wavelength of 3.5 m and a comparatively low active input impedance at a wavelength of 7 m. At a wavelength of 3.5 m, the high impedance from each dipole is scaled to the input of the twin-wire feeder through a whole number of half-waves without any transformation. The scaled input impedances of ten dipoles at the input of the span give a total active component of about 120 Ω and a small reactive component. At a wavelength of 7 m the low input impedance of each dipole is scaled initially through a quarter-wave KATV into a high impedance, and these impedances from the ten dipoles, scaled along the main feeder to the input of the span, when added together give an impedance with an active component also of about 120 Ω and a small reactance.

Two spans of the exciter are connected to a broadband symmetrizing device, in the construction of which ferrites are used. Calculation of the input impedance of these two spans of the exciter was carried out for frequency bands of 41-46 MHz and 82-88 MHz. The input impedances of the dipoles were taken from the experimental data. The results of calculations and experiments are given in Fig. 6, showing a good measure of agreement between the two. It will be noted that the calculated distribution of current along the dipoles was counted as uniform and in-phase. After adjustment of the additional capacitors at the input of the symmetrizing device to the working frequencies, their reflection coefficient does not exceed 0.15-0.2.

As mentioned previously, for the antenna at the out-station in Kalinin Region the diagram has to be inclined by 2.5° to the east of the in-phase position. Under these circumstances, if one span of the ten dipoles remains in-phase, the effective area of the antenna is reduced by about 35%, because a considerable diffraction effect appears. For this reason, an in-phase group of five dipoles was left for this antenna. An additional twin-wire feeder was mounted into each span, and this also was connected to a symmetrical screened twin-wire line with $w = 150\ \Omega$.

Two spans are joined into a section by means of a summing device. The input impedances of one span were calculated and measured. Agreement between the calculated and experimental data was satisfactory.

Effective Area of the Antenna. Calculations of the effective area of the antenna were carried out at a wavelength of 3.5 m. Values of the efficiency, taken in succession for the sections of the feeder circuit, are given below:

Dipole	0.98
KATV	0.92
Twin-wire feeder	0.97
Symmetrical screened line (150 Ω)	0.97
Symmetrizing device	0.94
Because of matching of section with 75 Ω circuit	0.975
RKM-5/18 cable, 50 m	0.82
First doublet ring	0.94
RKM-5/18 cable, 80 m	0.73
Second doublet ring	0.94
Total calculated	0.40
Total experimental	0.50 ± 0.05

These data show that the calculated efficiency of the dipole-feeder system is 0.40. Values of the efficiency of the dipole-feeder system were measured for both extension antennas on a wavelength of 3.5 m, for emission of the background component. For both antennas an experimental value of 0.5 ± 0.05 was obtained for the efficiency, in good agreement with the calculated value. Values determining the coefficient of surface use (CSU) are given below:

Because of leakage through reflector	0.95
Irregular amplitude distribution along dipoles in span	0.98
Phase scatter along dipoles	0.95
Spillover	0.71
Shadowing by subreflector	0.91
Irregular amplitude distribution at antenna aperture	0.92
Total	0.52

Finally, allowing for the efficiency and the total CSU of the dipole-feeder system for an antenna with geometrical area of 5600 m², we obtain the calculated value of the effective area of the antenna for a wavelength of 3.5 m, equal to 1160 m². The calculated value of the effective area for a wavelength of 7 m is about 1500 m².

TABLE 1.

Section No.	Antenna at Staritsa out-station							Antenna at Pereslavl'-Zalesskii out-station						
	1	2	3	4	1,2	3,4	1, 2, 3, 4	1	2	3	4	1,2	3,4	1, 2, 3, 4,
A_{eff}, m^2	—	—	—	228	656	512	910	277	320	320	300	650	593	—
$2\theta_{0.5\,p}$	—	—	—	2°29'	1°10'	1°16'	47'	2°29'	2°34'	2°36'	2°48'	1°06'	1°10'	34'
$2\theta_{0.0\,p}$	—	—	—	5°32'	2°56'	2°56'	1°27'	5°30'	5°45'	5°35'	5°45'	2°52'	2°54'	1°29'

Measurements of polar diagrams were made for separate sections of each half of both antennas and for the whole antenna for radiation from the radio source Cassiopeia A. The results of these measurements are given in Table 1. At a wavelength of 7 m the measured effective area of the antenna is about 1800 m^2.

To conclude, it must be noted that: (1) the antenna may work on two multiple waves, it has good directivity, and it has a large enough effective area to enable flickering of radio sources to be observed; (2) the dipole-feeder system is simple to adjust, reliable, dismountable, and its separate components can easily be replaced.

3. Radiometers in the 83-89 MHz Frequency Band

General Description. The radiometer is designed for the investigation of cosmic radio waves in the 83-89 MHz range. For studying solar wind, radio signals are recorded at three points. Two out-stations operate in conjunction with the cruciform DKR-1000 radiotelescope.

The radiometer consists of a two-channel radio receiver, based on a superheterodyne circuit, with double frequency conversion and a correlation output.

The weak noise signal of the radio source travels from the antenna to the input of the receiver channel along a radio-frequency coaxial cable with a wave impedance of 75 Ω. In the correlation method adopted for reception, signals reach each channel from the corresponding part of the antenna system.

The hf input amplifier is assembled from a bantam series of vacuum tubes. From the point of view of increasing the sensitivity of the radiometer it is important that the input circuit of the amplifier have a low noise factor and a reasonably high nominal power-amplification factor. The best of these tube circuits is the combination: triode with common cathode — triode with common grid. The input cascade of the amplifier is assembled with 6S3P and 6S4P tubes with an autotransformer input circuit.

Amplification of noise signals to the level required for normal operation of the correlator circuit is carried out in two if amplifiers. Both the first and the second of these amplifiers are assembled with type P403 transistors. The first if amplifier consists of three cascades, two of which are symmetrically detuned relative to a mean frequency of 30 MHz, while the second is built according to a video amplifier design. The voltage of the heterodyne frequency is fed into the mixers of the separate channels through decoupling buffer cascades. To maintain the phase relationships of the signals used, a common heterodyne for the two channels is applied to the inputs of the multiplier circuit. The necessary reception band is provided by installing highly selective filters after the second mixer. A special feature of this radiometer is the use of three interchangeable bands (1.5, 0.5, 0.1 MHz), with the possibility of scanning the reception band within the limits of the band of the first if amplifier. The scanning band is about ±3 MHz.

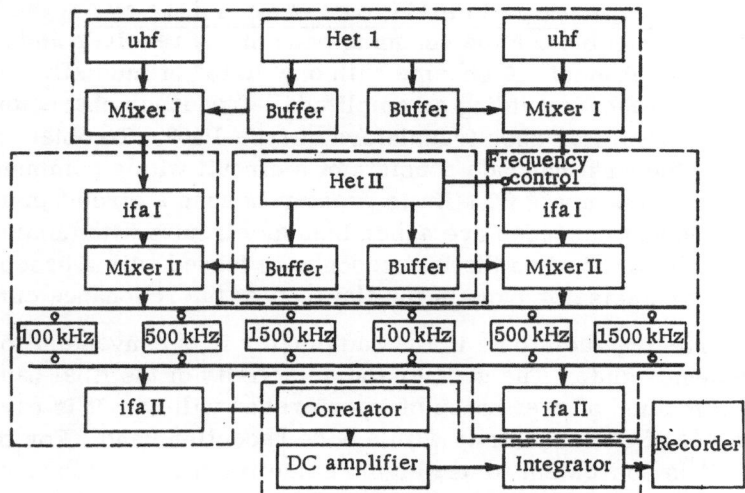

Fig. 7. Block diagram of the radiometer.

After multiplication in the correlator circuit, the dc signal, proportional to the product of the input signals, is amplified in the lf system and integrated. The dc amplification is effected by a circuit with amplitude modulation of the input signal. Transistors of types P13 and P15 are used in the low-frequency part of the radiometer. A block diagram of the radiometer is given in Fig. 7.

The radiometer is assembled as separate screened subchassis. The three subchassis are: the radio-frequency unit, intermediate-frequency unit, and low-frequency unit and correlator. The units are connected together by matched coaxial cables. This type of construction makes it possible to replace any of the subchassis quickly in case of breakdown of any part of the circuit or if certain modifications become necessary. It also enables rapid inspection to be made and the location of a fault can be conveniently sought; the individual units of the circuit are well screened from each other. All the subchassis are mounted on one common chassis, where the regulating elements, controls, and monitoring instruments are housed.

The semiconductor circuits are housed in heat-insulated jackets; foam polystyrene is used as the insulating material. The semiconductor circuits are fed by a stabilized power source with output voltage $E = 12$ V and a stabilization factor of more than 4000. The tube part is powered from standard sources: a stabilized anode voltage $E_a = 150$ V and a filament voltage $E_f = 6.3$ V, with a stabilization factor of about 200. Let us now examine some of these circuits in more detail.

High-Frequency Amplifier. The choice of type of amplifier is determined by the desire to obtain a minimal noise factor in the first cascade with the maximal nominal power amplification. At a frequency of 86 MHz it is best to use a mixed cathode — grid circuit. In the receiver the circuit is built with 6S3P and 6S4P triodes. The input system is designed as an autotransformer circuit with transformation factor providing a minimal noise factor to the amplifier. A resonance system with two pairs of circuits mutually staggered at the critical frequency difference is used to obtain the required frequency characteristic. After the "cascode," cascades with grounded cathodes, incorporating 6Zh1P tubes, are used.

The use of a circuit "in pairs" increases the product of the transmission band and the amplification factor of the cascade by 1.5 times compared with the design with circuits tuned to one frequency, and also gives a resultant resonance curve of better shape. With a band of $P_{0.7} = 8$-10 MHz and a mean frequency $f_{mean} = 86$ MHz, the amplifier gives a noise factor $N = 1.8$-1.9 and an amplification factor $k_0 = 400$.

Intermediate-Frequency Amplifiers and Filters. The first intermediate-frequency amplifier is responsible for the scanning band of the receiver and for weakening of the signal along the mirror channel. A scheme with one "trio" of mutually staggered circuits is chosen, the frequency difference providing an amplitude — frequency characteristic with the flattest possible curve. The amplifier uses transistors of type P403. To obtain the smallest noise factor for the amplifier, the first cascade consists of a circuit with a common base, because power amplification at a frequency of 30 MHz is greater in such a circuit than in one with a common emitter. The subsequent cascades are assembled as circuits with common emitters. The amplification factor of the "trio" when the last cascade is loaded with a cascode frequency changer is 10–15. The transmission band in the flat part of the resonance curve is 6 MHz.

To get rid of interfering radiations in the radiometer it is convenient to use a predetermined heterodyne frequency shift. The greater the selectivity of the final band-pass amplifier, the more effective this method of prevention of interference will be. It is particularly important to have a good coefficient of rectangularity with a wide reception band. For the values of bands required, in this case it is convenient to use highly selective filters, incorporated as loads in the second mixer. Interchangeable band-pass filters with a simple chain circuit are used in the design of the receiver.

The second intermediate-frequency amplifier provides for the chief amplification of the signal to a value essential for normal operation of the multiplier circuit. The amplifier must have a factor $k_0 \approx (5-10) \cdot 10^3$ and a uniform frequency characteristic curve in the widest reception band. To satisfy these demands, the amplifier is built with type P403 transistors using the design of a video amplifier circuit without correction. The last cascade of the amplifier has a transformer load; the output transformer is used in the multiplying circuit.

Direct Current Amplifier. The main disadvantages of dc amplifiers are their temperature and time drifts and also the fact that, because of galvanic connections in the circuit, the whole spectrum of low-frequency noise is amplified and passes unhindered to the output. This is particularly serious for amplifiers incorporating transistors.

These difficulties are largely overcome in dc amplifiers with modulation of the input signal. The dc signal in this arrangement is first converted into an ac signal proportional to it (by means of a modulator), and then amplified. Later (depending on the character of processing of the signal after amplification) it can be transformed into a dc signal by means of a linear detector incorporating a transistor. The main problem arising during this method of amplification is the stability of the modulator itself. Meanwhile, a source of modulating voltage must be provided in the circuit.

A transistorized modulator is used in the circuit of the dc amplifier. It consists of a half-wave converter, enabling exact reversals of the voltages to be carried out. The circuit for exact reversing of the voltages is based on the fact that a transistor, operating under conditions of deep saturation, is stable. During modulation of low levels of power, the factors limiting the use of such converters are the residual voltage and current, the drift of these values with a change in temperature, and the time required for the transistor to change from one state to another. The converter in the radiometer is built with two type P15 transistors connected in inversion. The opposed connection of the two transistors enables an identity of reversal of about 0.05% to be obtained. Such a switch possesses two-way conductivity, and ensures mutual compensation of the residual voltages and currents and of their drift during a change in temperature, for they act in opposition. The amplitudes of the overshoots during reversal also are reduced. The modulating voltage is supplied by a transformer, and the internal resistance of the generator of the modulating voltage must be low. The secondary winding in the circuit is shunted by a 150-Ω resistor. The switch changes the load on the correlator with the frequency of modulation within a range from the nominal value to zero. The degree of modulation under

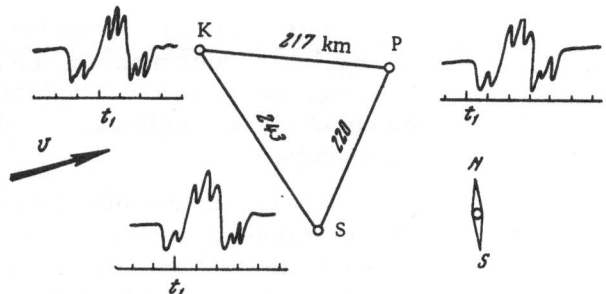

Fig. 8. Diagram showing relative position of three observation points and showing recordings from them with
time shifts attributable to solar wind blowing in the direction
indicated by arrow. K) Kalinin; P) Pereslavl'; S) Serpukhov.

these circumstances is close to 100%. During application of the modulating voltage U_{mod} =
300-600 mV and compensation effected by selection of basic resistances of the transistors used
in the switch, it is possible to obtain a parasitic signal of the order of 10-25 μV, brought to the
output of the circuit, and a coefficient of transmission of about unity.

This transistorized modulator has essential advantages over the contact type, the working
frequency of which is limited, its length of service short, its reliability insufficient, and its
screening cumbersome (because of the presence of inductance from the coil of the electromagnet).

The source of modulating voltage with a frequency of the order of 1 kHz is a generator
with a resonance circuit in the collector circuit. The generator incorporates a type MP-39
transistor; with a current of 7 mA it provides a voltage of 800 mV at the output.

To amplify the ac signal proportionally to the amplitude of the dc input, a low-frequency
amplifier is used, based on the circuit of a resonance RC-amplifier, in which a twin T-shaped
bridge serves as the frequency-selecting feedback circuit.

All the circuits described are similar in assembly and control. Servicing of the radiometer is carried out on the basis of general methods of working with radio equipment.

4. Radio-Astronomical Observations of Solar Wind

Organization of Synchronous Radio-Astronomical Observations at
Three Points. Observations at points in the neighborhood of the towns of Serpukhov,
Kalinin, and Pereslavl'-Zalesskii were begun in June, 1966 and continued for almost two months.
The relative position of the three observation points and a scheme of the recordings are shown
in Fig. 8. The radio telescopes described earlier were set up at the out-stations. Great attention was paid to coordination of the recordings in time. Synchronization was carried out by
means of accurate time signals transmitted on frequencies of 5 and 10 MHz by radio stations of
the time service.

Time signals were fed every second from the output of a coupled receiver to an electronic
relay, controlling a mechanical relay, the armature of which was connected to the pen of an
automatic writer. When the relay is operated by the time signals the pen makes a short mark.
Because of the rapid winding of the graph paper, the distance between the seconds marks is 6
mm. The tested stability of the paper winding speed, affecting the accuracy of the readings
under the conditions of the DKR-1000 mobile laboratory, proved to be quite satisfactory. In 20
measurements during a work period of 1.5 h (of the automatic writer), dispersion in determination of a time interval of 1 sec was 0.0009 sec (~0.1%). The construction of the antenna and
radiometers has been fully described above.

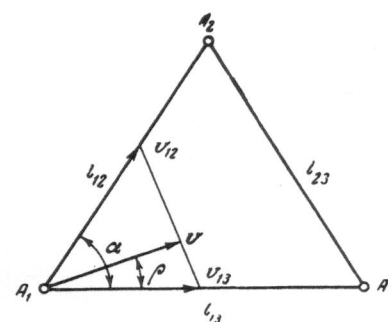

Fig. 9. Diagram of location of antennas. A_1, A_2, A_3) Reception points; v_{12}, v_{13}) apparent velocities; v) velocity and direction of movement.

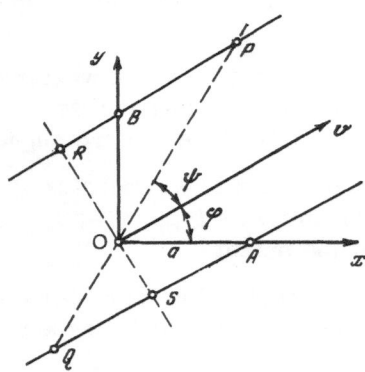

Fig. 10. Diagram of location of reception points. O, A, B) Observation points; v) velocity of movement of diffraction pattern; φ) angle between direction of velocity and base OA; ψ) angle between direction of velocity and "line of maximum amplitude."

Analysis of Results of the Observations.

To determine the velocity of the movements, it is clearly necessary to analyze the curves of measurements of signal amplitude obtained at the three reception points.

Let us consider how, by investigating the character of fluctuations of signal amplitude at three points located at the apices of a triangle, the velocity of drift of the diffraction pattern near the earth and the dimensions, shape, and orientation of irregularities in the diffraction pattern can be determined. The method is based on analysis of the overall similarity between the three curves. Having determined to what extent they differ from one another by comparison with what would be obtained in the case of an unchanged diffraction pattern, the character of its change can then be established. The results obtained by this analysis will relate only to the characteristics of the diffraction pattern on the earth's surface. We are not concerned at this stage with the relationship between this pattern and the phenomena in the solar wind producing it. The question of any particular model of the solar wind does not enter into the discussion.

The character of the diffraction pattern can be represented by lines of constant amplitude, and in future we shall speak of the shape, dimensions, and direction of its movement, referring to contours of constant amplitude. How are the principal parameters found? Correlation analysis is the usual method. A series of maxima and minima of each of the three recordings is taken and mathematical analysis of the results of the observations is carried out. However, because of the impossibility of computer analysis in our case, we have not yet used this method. We have chosen the "similarity method." Only obviously similar parts of the curves obtained at the three points are analyzed. The time shifts are determined at characteristic points on these curves, such as the maxima and minima of fluctuations of amplitude. Then, knowing the geometry of the arrangement of the reception points, the velocity and direction of movement of the diffraction pattern are determined. This method of analysis of the results of the observations has been widely used with measurements of the velocities of movement of ionospheric irregularities.

The velocity and direction of the movements in this case are determined by the following equations [10]:

$$v = \frac{l_{13}}{t_{13}} \cos \beta = \frac{l_{12}}{t_{12}} \cos (\alpha - \beta),$$

$$\tan \beta = \frac{\dfrac{t_{12} l_{13}}{t_{13} l_{12}} - \cos \alpha}{\sin \alpha},$$

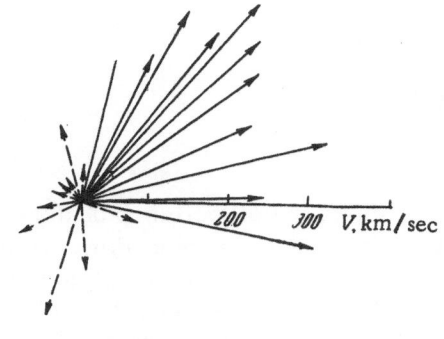

Fig. 11. Results of determination of velocity of interplanetary plasma obtained by the method of similarity (not all results obtained are plotted).

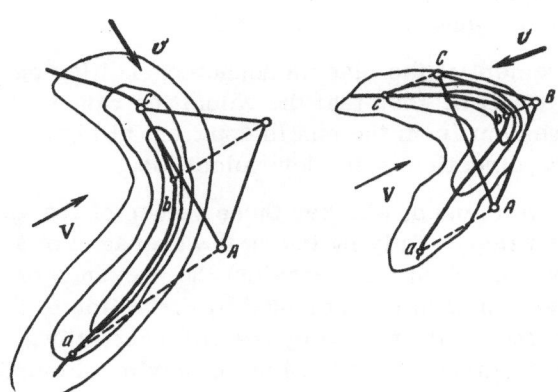

Fig. 12. Examples of possible forms of diffraction details leading to considerable deviations of the observed direction of movement of v from the true V. V represents the true velocity, v the measured (apparent) velocity.

where $l_{12}/t_{12} = v_{12}$ and $l_{13}/t_{13} = v_{13}$ represent the apparent velocities in the directions A_1A_2 and A_1A_3, respectively (see caption to Fig. 9); A_1, A_2, and A_3 the reception points; t_{12} and t_{13} the time of delay of the diffraction pattern between the corresponding points; l_{12} and l_{13} the distance between the points.

An attractive feature of this method of similarity is its simplicity. One of its drawbacks is its low efficiency, for only curves which are clearly similar can be used in the analysis. A more serious disadvantage, however, is the low accuracy of determination of the velocity and direction, and the complexity of analysis of the anisotropy of shape of the diffraction pattern. The possibility of a considerable difference between the direction of true movement and the movement obtained by the method described, because of the complex geometry, must be borne in mind.

Let us examine a simple case, as is done in [11]. Let the receivers be located at the points O, A, and B (Fig. 10), where O is the origin of the coordinates, A the point (a, O), and B the point (O, a).

Let the unchanged diffraction pattern move with velocity v in the direction forming the angle φ with Ox, and let part of the "line of maximum amplitude" be represented by the intercept OP, forming the angle ψ with the direction of the velocity of drift v. Let OP pass through O at the moment of time t, through A at time $t + t_x$, and through B at time $t + t_y$.

$$t_x = \frac{QS + AS}{v} = \frac{a}{v}(\sin\varphi \cot\psi +$$

$$+ \cos\varphi) = \frac{a}{v}\cos\varphi(1 + \tan\varphi \cot\psi), \qquad (1)$$

$$t_y = -\frac{(RP - RB)}{v} = -\frac{a}{v}(\cos\varphi \cot\psi - \sin\varphi) = \frac{a}{v}\sin\varphi(1 - \cot\varphi \cot\psi). \qquad (2)$$

Let us assume that the velocity and direction of the wind are constant (i.e., that v = const, φ = const), and that the angle ψ assumes all values arbitrarily, i.e., corresponds to a random orientation of the lines of maximum amplitude. If, then, a sufficient number of time shifts t_x and t_y is taken, and they are averaged, we obtain

$$\bar{t}_x = \frac{a}{v}\cos\varphi, \qquad (3)$$

$$\bar{t}_y = \frac{a}{v} \sin \varphi, \qquad\qquad\qquad (4)$$

$$(\bar{t}_x)^2 + (\bar{t}_y)^2 = \frac{a^2}{v^2}, \qquad\qquad\qquad (5)$$

$$\bar{t}_y / \bar{t}_x = \tan \varphi. \qquad\qquad\qquad (6)$$

These results show that the value of the velocity v and the direction φ of the constant wind can be deduced from the mean time delays ($\bar{t}_x$ and $\bar{t}_y$) taken from the recording. In the case examined here, the receivers are located at the apices of a right-angled triangle with legs equal to a. As was mentioned earlier, the reception stations built for measuring the velocity of the solar wind are located at the apices of an equilateral triangle with sides about 220 km in length.

During the period of the observations (July — August, 1966) many recordings were obtained from the sources 3C-144 and 3C-147. Recordings from the three points were compared relative to their synchronizing marks, parts of the curves showing a sufficient degree of similarity were chosen, and from the position of characteristic points (maximums or minimums of the fluctuations) the relative time shifts were determined. From the known geometry of the points and the obtained values of the time shifts, the directions and velocities of movement of the diffraction pattern were determined. Assuming the velocity of the diffraction pattern to be equal to the velocity of the nonhomogeneities in the solar wind, and projecting these velocities on the celestial sphere, the pattern of movement of the nonhomogeneities can be obtained.

The values of the velocity and direction of movement of the plasma nonhomogeneities determined as described above are shown in Fig. 11. It will be seen that the velocities range from 350 to 20 km/sec. The radial direction of movement from the sun is apparent at high velocities, and it is impossible to discern any definite direction of the low velocities.

It is difficult to say, when this method of analysis is used, whether these values of the apparent velocities and directions are close to the true values. Only by taking the averages of a large number of measurements can the correct answer be obtained. Possibly the presence of very low values of the velocities and of different directions can be explained by distortion of the shape of the lines of maximum amplitude of the diffraction pattern and by the different values of the angles between the direction of movement and the direction of the line of maximum amplitude. Under these circumstances completely opposite directions and substantially underestimated values of the velocities may be obtained (Fig. 12).

Remembering that other methods of studying the dynamics of interplanetary plasma give near-radial direction of travel from the sun, and that there are fewer cases of velocity directed toward the sun (only four points in one recording) than away from it, it is reasonable not to consider them yet. If, then, it is assumed that the magnitude and direction of the velocity are constant, and only the direction of the lines of maximum amplitude of the diffraction pattern relative to the direction of the velocity varies, Eqs. (5) and (6) can be used to determine the direction and velocity of movement of the diffraction pattern. In this way mean velocities of about 250 km/sec are obtained, if only data close to radial are analyzed, and lower values if all the data are averaged.

LITERATURE CITED

1.   V. V. Vitkevich, Dokl. Akad. Nauk SSSR, Vol. 77, No. 4 (1951).
2.   V. V. Vitkevich, Dokl. Akad. Nauk SSSR, Vol. 101, No. 3 (1955).
3.   A. Hewish, Proc. Roy. Soc., 228:239 (1955).
4.   V. V. Vitkevich and B. N. Panovkin, Astron. Zh., Vol. 36, No. 3 (1959).
5.   V. V. Vitkevich, Astron. Zh., Vol. 35, No. 1 (1958).

6.   O. B. Slee, Australian J. Phys., Vol. 12, No. 2 (1959).
7.   V. V. Vitkevich, V. I. Babii, and A. G. Sukhovei, Astron. Zh., Vol. 42, No. 1 (1965).
8.   V. F. Bedevkin and V. V. Vitkevich, in: Wideband Cruciform Radio Telescope Research
     (D. V. Skobel'tsyn, ed.), Consultants Bureau, New York (1969), p. 86.
9.   S. N. Ivanov and Yu. P. Ilyasov (in press).
10. V. V. Vitkevich and Yu. L. Kokurin, Radiotekhn. i Elektron., Vol. 4, No. 1 (1959).
11. J. A. Ratcliff, J. Atmospheric Terrest. Phys., 5:173 (1954).

RADIO-RELAY INTERFEROMETER
FOR THE METER WAVEBAND

G. I. Dobysh

An interferometer with radio relay has now been produced at the Radio Astronomy Station, Physical Institute of the Academy of Sciences, for preliminary trials. The suggested maximum base length is 40-50 km.

For various reasons the frequencies of radio waves from sources selected for reception were within the 34-36 MHz band (λ = 8.3-8.8 m). The fact that in this band there is a comparatively high level of atmospheric and industrial interference largely predetermined the methods and techniques used in individual units and components of the apparatus. One of the chief conditions was the ability to readjust the radio-astronomical receivers smoothly within the range of 1-2 MHz so as to enable areas with the lowest noise level to be chosen within the stipulated waveband.

The general principles of construction of interferometers and individual results are well known from previous publications [1-4]. However, practical planning difficulties have been due to the absence of concrete parameters for several of the chief parameters of these systems, and above all for the relay apparatus. In particular, the following parameters were unknown: (1) permissible levels of noise, background, and stray radiations within the band of working frequencies; (2) permissible coefficients of nonlinear distortion; (3) inequality of the group time in the hf circuit; etc.

There were also unsolved problems connected with the formation of coherent heterodyne frequencies for spaced radio-astronomical receivers. As a basis for development of the apparatus, the recommendations of the International Consultative Committee on Radiocommunication (ICCR) for radio-relay lines with frequency compression were adopted. However, the parameters and technical conditions can be finally specified and their details clarified only after experimental verification. This particular apparatus was planned and executed with this possibility in mind.

This type of interferometer consists of two spaced receivers of radio waves from cosmic sources (radio-astronomical receivers), with a coherent heterodyne. This common heterodyne for both receivers is located at a distant point (DP) and is transmitted along a radio-relay line to the principal point (PP) together with the radio-noise signal. Relaying of the radio-noise signal and transmission of the heterodyne frequency are performed by the same transmitter. Reception of these two signals at the PP is also carried out by one receiver, with subsequent separation after demodulation and amplification.

The interferometer instrument consists of the following principal units:

a) At the distant point: a radio-astronomical receiver, radio-relay transmitter with frequency modulation (FM), stabilized power sources, and control systems;

b) At the principal point: a radio-astronomical receiver, a receiver of relayed signals, a balancing unit and filters, delay line, low-frequency part, stabilized power sources, and control system.

Block diagrams of the DP and PP are given in Figs. 1 and 2, respectively.

The radio-astronomical receivers are built with a superheterodyne design with heterodyne frequency located in the middle of the transmission band of the high-frequency (uhf) amplifier, i.e., the nominal value of the heterodyne frequency f_{het} in this particular case is given by the equation $f_{het} = (f_h - f_l)/2$, where f_l and f_h represent the frequencies at the limits of the uhf transmission band of the receiver. This method of transformation was used for the following reasons: (1) the transmission band of the intermediate frequency amplifier (ifa) is narrowed by half since the effective frequency band, determining the limiting sensitivity of the interferometer, is in this case equal to $2\Delta f_{ifa}$; (2) relaying of the local-frequency spectrum of the zero-frequency signal to some extent simplifies the transmitter, reduces the demands on several characteristics, and enables it to be used (in spite of its comparatively narrow band) for simultaneous transmission of the heterodyne frequency and noise signal; and (3) transfer of the noise-signal spectrum into the low-frequency region is also simplified by the provision of a delay line, essential in this case for compensating the time of spread of one noise signal around the path of the radio-relay line.

The high-frequency part of the receiver consists of a preamplifier, located immediately next to the antenna, and the uhf receiver proper, assembled as two amplifiers connected in parallel, either of which can be connected at will, and having transmission bands of 1 MHz and tuned bands of $f_0 = 34.5$ MHz and $f_0 = 35.5$ MHz, respectively. The antenna amplifier has a transmission band of 6 MHz with flat apex in the region 35 ± 1.5 MHz and a noise factor N = 3.85. Smooth tuning of the receiver is thus possible between 34 and 36 MHz without any retuning of the uhf receiver, and the selectivity of the input part is sufficiently high.

Next the signal is transformed into the intermediate frequency in a mixer by means of the heterodyne frequency $f_{het} = 35 \pm 1.5$ MHz. The ifa consists essentially of a low-frequency amplifier with high negative feedback ($k\beta = 120$), the transmission band of which is from 1 to 210 kHz. The frequency characteristic of the receiver in the intermediate frequency is actually formed in the region of low frequencies by intermediate circuits, but in the region of high frequencies by a three-component low-frequency filter with cutoff frequency of 135 kHz, coupled after the mixer and creating an attenuation of 60 dB at a frequency of 180 kHz. After the filter, a stepwise amplification control with range of up to 30 dB is connected to the input of the ifa.

Because of the strong negative feedback, during a change in the supply voltage (at the anode or filament) of $\pm 20\%$ the amplification factor at the intermediate frequency is changed by less than 1%. Since in the ifa there are no resonance circuits, the shape of its frequency characteristic and its transmission band remain unchanged for changes of the tubes and regulation of the amplification.

The voltage of the heterodyne with frequency 35 ± 1.5 MHz at the distant point is obtained by multiplying the generator frequency tuned at between 558 and 608 kHz, by 60 times. The tuned generator is so designed that, with suitable thermostabilization, a long-term stability of $1 \cdot 10^{-4}$ can be obtained.

The amplitude of the heterodyne voltage at the mixer is given by $U_{het} = 12$ V. The coefficient of transformation thus remains constant during changes in U_{het}, at least by $\pm 30\%$. The effect of amplitude in stability of the heterodyne multiplier unit on the amplification factor of

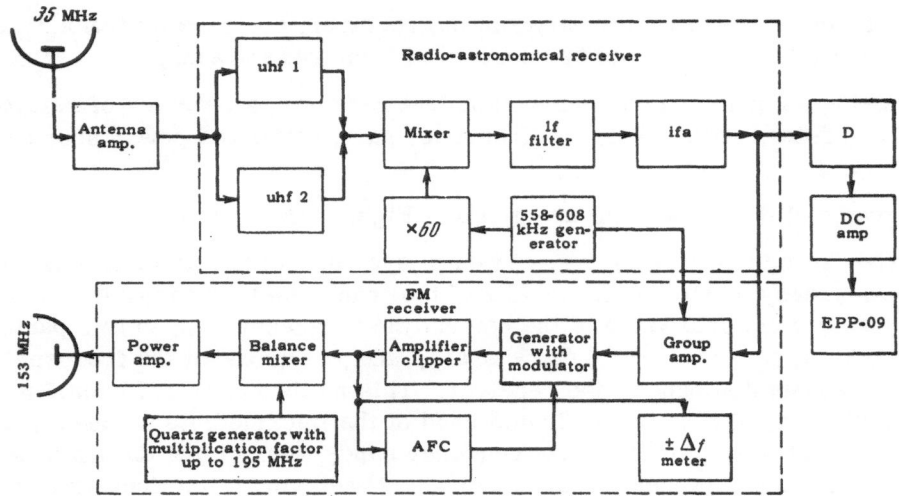

Fig. 1. Block diagram of apparatus at the distant point.

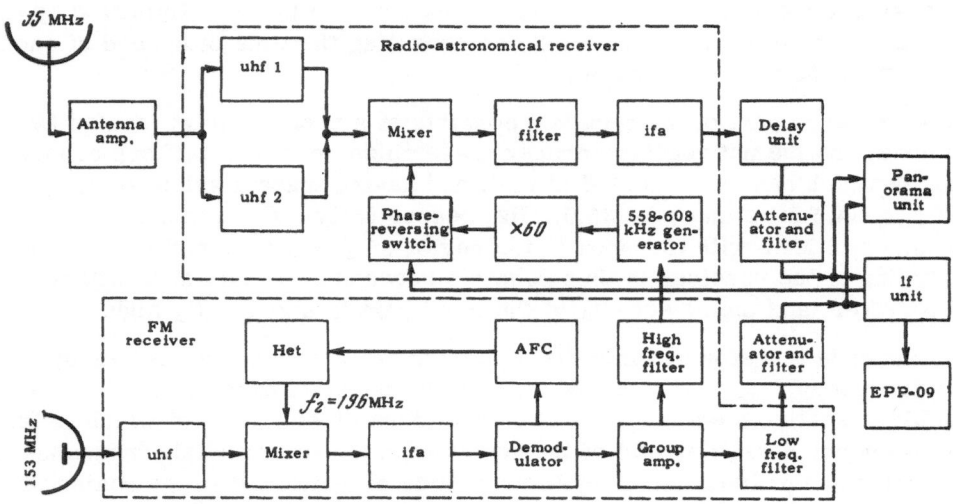

Fig. 2. Block diagram of apparatus at the principal point.

the radio-astronomical receiver is thus virtually eliminated. The noise signal with a spectrum of between 1 and 135 kHz and part of the voltage of the 558–608 kHz generator are fed from the output of the radio-astronomical receiver at the DP into the modulator of the radio-relay line transmitter.

The radio-relay line transmitter operates on a frequency of 153 MHz. Its maximum output power is 50 W. It is frequency modulated. The relay frequency of 153 MHz was chosen on the grounds that in the meter waveband there are no rapid and deep fadings of the signal along their path due to reflections from nonhomogeneities in the troposphere and changes in refraction. In its simplified form, the change of phases between the direct and reflected waves because of the difference in path at wavelength λ can be determined by

$$\Delta\varphi = \frac{2\pi}{\lambda}\,\Delta r\,(g),$$

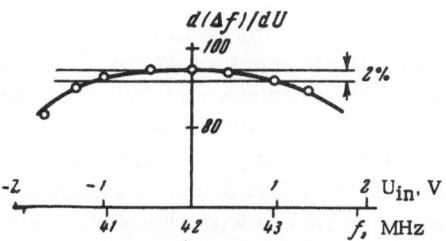

Fig. 3. Gradient of modulation characteristic curve of transmitter modulator as a function of frequency.

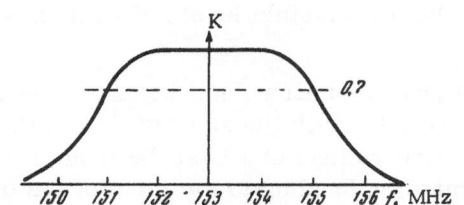

Fig. 4. Transfer function of transmitter.

where $\Delta r\,(g)$ represents the change in the difference of path as a function depending on the vertical gradient of dielectrical permeability of air g, the character of the path, etc. For the middle zone of the Soviet Union, and for a path R = 45 km, the value of $\Delta r\,(g)$ will be of the order of 12 cm. In that case, for λ = 10 cm, $\Delta\varphi$ = 432°, and during a change in g a maximum and minimum of interference may appear, i.e., at the receiving point the field intensity will undergo severe changes. At λ = 2 m, $\Delta\varphi$ is only 21.6°, and the signal level at the receiving point will be much less dependent on the conditions of reception.

To determine the output power of the transmitter, besides taking all other factors into consideration, the possibility of working on comparatively simple antennas (for example, of the "waveguide" type), which are convenient for transportation and for installation on light supports, served as the initial basis for the discussion. If the power of the transmitter is 50 W, then for a base measuring R = 45 km, the required amplification factor of the antennas of the radio-relay line of communication will be G = 16 dB. The possibility of a smooth change in output power of the transmitter within the range from 5 to 50 W was provided for.

As was pointed out above, FM is used in the transmitter. The advantage of FM over AM is obvious, and it lies above all in the possibility of virtually completely eliminating the stray amplitude modulation due to fadings along the path and inequality of the frequency characteristics of the hf circuits of transmitter and receiver. The FM modulator is constructed of parallel circuits [5]. Because of this design, a high degree of linearity of the modulation characteristic can be obtained over high deviations of frequency.

The central frequency of the modulator f_0 = 42 MHz, and the maximum deviation of frequency Δf_{max} = 1.2 MHz. The relationship obtained by means of a model between the gradient of the modulation characteristic $d(\Delta f)/dU$ and the input voltage of the modulator is shown in Fig. 3. It is clear that deviations from a linear law do not exceed 2%, corresponding to a coefficient of nonlinearity in the second and third harmonics of the order of 0.2%. With more careful control it is possible to obtain a coefficient of nonlinear distortions less than 0.1%. The demand for a high degree of linearity of the modulation characteristic curve is explained primarily by the two-signal modulation of the transmitter, for otherwise mutually crossing distortions and interference would be possible in the heterodyne and noise channels.

After amplification and clipping, the FM waves with a central frequency of f_0 = 42 MHz and peak deviation Δf_{peak} = ± 830 kHz (±360 kHz in the noise channel and ±470 kHz in the heterodyne channel) are fed into the balance mixer, where they are mixed with a frequency f_{het} = 195 MHz, obtained from a quartz generator after multiplication. The difference frequency

$$f_{het} - (f_0 \pm \Delta f_{peak}) = 153 \pm 0.86 \text{ MHz}$$

is further increased to 50 W by a two-cascade power amplifier, built around ultrashort-wave generator tetrodes of the GU-29 type. The transfer function of the transmitter is shown in Fig. 4.

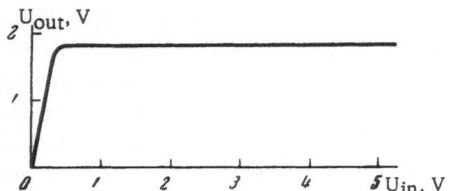

Fig. 5. Characteristic curve of clipper of FM radio-relay receiver at the principal point.

By means of the automatic frequency control (AFC) of the central frequency at 42 MHz, and the quartz generator for a frequency of 195 MHz, stability of the central frequency of the transmitter at $2 \cdot 10^{-5}$ was obtained. During operation, continuous control over deviation of frequency is effected by the deviation monitor linked with the AFC system.

The following fact must be noted. Experiments have shown that the presence of combination and accessory frequencies in the band of emission of the transmitter with levels of 60-65 dB from that of the central wave without modulation, which is in accordance with the ICCR norms, is quite unacceptable because it may lead to the appearance of interference in the channel of the PP radio-astronomical receiver by the channel of the relayed heterodyne frequency. Measurements showed that the permissible levels of combination frequencies are 80-90 dB.

The radio-relay line receiver, located at PP, has one uhf cascade, a mixer, a heterodyne with AFC system, a 6-cascade ifa, a demodulator, and a group lfa with linear cascades at the output. The sensitivity of the receiver is 30 μV for an effective voltage of 1 V at the input of the demodulator clippers. The noise factor N = 3. The transmission band up to the input of the demodulator is 4.7 MHz.

In FM, the nominal width of the frequency band of the system is given approximately by:

$$2\Delta f = 2 \left(\Delta f_{max} + F_{max} \right),$$

where Δf_{peak} represents the peak deviation of frequency, and F_{max} the highest frequency of the spectrum of the modulating signal. Because of inaccuracy of tuning of the receiver, instability and irregularity of the uhf and ifa characteristics, instability of the frequency of the transmitter and heterodyne of the receiver, and various other factors, the band width must be increased by about 1.6 times. The final values of the FM band of the receiver are thus given by the expression

$$3.2 \left(\Delta f_{peak} + F_{max} \right),$$

and in the case under discussion we obtain $2\Delta f = 4.6$ MHz. For stabilization of the frequency characteristic and, what is particularly important to FM, the phase characteristic, overall values of capacitances for the ifa circuits were selected to be of the order of 50 pF. Because of this measure, the shape of the frequency and phase characteristic curves of the ifa are independent of changes in the operating conditions, or aging and replacement of the tubes.

To overcome distortions because of multiple-wave reception and attenuations on the path in the demodulator, two-cascade broad-band clipping was used. The characteristics of the clipper are shown in Fig. 5, where the point $U_{in} = 1$ V corresponds to a signal of 30 μV at the input of the receiver. A clipping level of 1.8 V at the input of the discriminator is maintained with a high degree of accuracy and stability by using limiting diodes of dry cells of the "Saturn" or "FBS" type as the sources of grid bias. Since the anticipated changes in signal level due to fading for a path R = 45 km and λ = 2 m do not exceed 11 dB, corresponding to a change of 3.55 times in E_{in} of the receiver, during 99.9% of the time, they will be completely suppressed, so that the signal level at the input of the discriminator remains constant during changes in the signal at the clipper input (counting from a threshold level $U_{in} = 1$ V) by at least 14 dB.

The discriminator operates on staggered tuning. The slope of the characteristic curve of the discriminator is 200 mV/MHz. Nonlinearity within the limits of ±2 MHz from the central frequency is the same as for the transmitter modulator.

After leaving the output of the demodulator, the noise and heterodyne signals are amplified in a group amplifier, covered by high negative feedback to a level of 0.5-1 V, after which the noise signal passes through a low-frequency filter (f_{mean} = 180 kHz) to the balancing unit and filters, while the heterodyne signal passes through a high-frequency filter (f_{mean} = 450 kHz) to the heterodyne frequency multiplying unit of the radio-astronomical receiver at PP.

In this unit the frequency of 558-608 kHz relayed from the DP is multiplied to a frequency of 35 ± 1.5 MHz (just as at the DP, the multiplication factor is 60), and after amplification it is used as the heterodyne voltage of the radio-astronomical receiver at PP.

Expected variations in phase of the heterodyne frequency at the principal point after multiplication to 35 ± 1.5 MHz, produced by a change in the path "length" because of refraction, have the value $\Delta \varphi$ = 2.9°, which is quite acceptable. As the first experiment showed, the main difficulty in direct multiplication of the relayed frequency of 583 ± 25 kHz is not merely reaching a high level of phase stability, but also obtaining a heterodyne voltage which is sufficiently free from noise, i.e., a heterodyne voltage at the mixer of the radio-astronomical receiver at PP with a high signal-to-noise ratio. For these purposes we now use synchronization of the local generator by tying its frequency to the frequency of the relayed voltage of 558-608 kHz. The synchronization procedure is controlled oscillographically. With sufficiently rigid phase synchronization it is still impossible in this case to obtain a high level of protection against interference in the heterodyne at PP, although it is significantly higher than when direct multiplication is used.

The next stage is to investigate the possibility of using a phase ifa system. The initial arguments are as follows. The multiplying unit to enable the heterodyne frequency to be changed has a transmission band of 0.1 MHz at the input (corresponding to 6 MHz at the output). Most cascades work under extremely nonlinear conditions. As a result of this, the signal-to-noise ratio at the output of the unit is 10-15 dB lower than at the input. Even a very slight noise and interference level within the limits of the input band of the multiplier is considerably above the intrinsic noise level of the radio-astronomical receiver at PP, since the noise components of the heterodyne voltage participate equally in the conversion process with the useful signal.

The most obvious and effective method of reducing the noise level is in this case to narrow the transmission band of the multiplier unit. However, the construction of a narrow-band multiplier system with a relatively high multiplication factor, which can be retuned in a certain frequency range, and still have stable phase characteristics, is an extremely difficult task.

The necessary ratio between the heterodyne voltage and its noise $(U_{het}/U_{n.het})_{out}$ at the output of the multiplier unit and input of the mixer of the radio-astronomical receiver can be found from the equation

$$\left(\frac{U_{het}}{U_{n.het}}\right)_{out} = 20\log\frac{U_{het}}{2K_{uhf}}\sqrt{\frac{P_{n.m}/P_{n.het}}{kT_0 2\Delta f_{uhf}} \frac{}{R_a(N-1)}} \ [dB], \tag{1}$$

where $P_{n.m.}$ represents the power of noise generated by the high-frequency part of the radio-astronomical receiver including the mixer, when the heterodyne voltage is free from noise; $P_{n.het}$ is the noise power of the heterodyne voltage; and K_{uhf} is the amplification factor relative to the voltage of the uhf receiver before the grid of the mixer. So as not to increase the noise factor of the radio-astronomical receiver considerably, the ratio $P_{n.m}/P_{n.het}$ must be fairly high. Assigning the value $P_{n.m.}/P_{n.het}$ = 10, in the present case we obtain $(U_{het}/U_{n.het})_{out}$ = 100 dB.

If allowance is now made for the decrease in signal-to-noise ratio on account of the multiplier by 15 dB, the ratio $(U_{het}/P_{n\,het})_{in}$ at the input of the multiplier unit must be $\geq$ 115 dB. This ratio must also hold good at the output of the FM receiver of the radio-relay line for the channel relaying the generator frequency of 558-608 kHz.

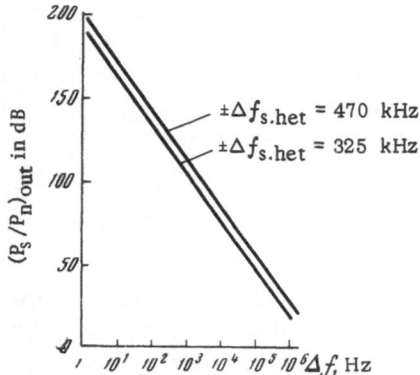

Fig. 6. Ratio P_s / P_n at output of LFA of radio-relay receiver as a function of transmission band of LFA.

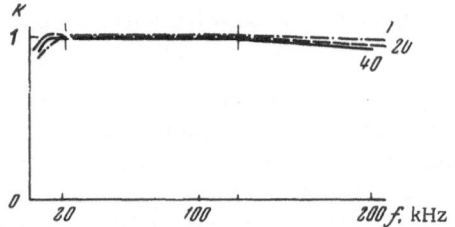

Fig. 7. Family of frequency characteristic curves of the delaying unit at different values of τ_d (values of delays indicated by numbers in μsec).

Using the results described in [6], we obtain the relationship between $(P_s/P_n)_{out}$ at the output of the FM receiver and $(P_s/P_n)_{in}$ at the input and the transmission band of the low-frequency part ΔF. In doing this, allowance is made for the fact that the deviation from the frequency of 558-608 kHz, $\Delta f_{s.het}$ is less than half the transmission band of the ifa of the receiver. In that case,

$$\left(\frac{P_s}{P_n}\right)_{out} = \frac{0.375\left(\frac{P_s}{P_n}\right)_{in} D^3}{B^2\left\{1 + 0.89\, D^2\left(\frac{P_s}{P_n}\right)_{in}\left[\exp\left(\frac{P_s}{P_n}\right)_{in} - 1\right]^{-1}\right\}},$$

(2)

where $D = 2\Delta f_{if}/\Delta F$ has the meaning of a coefficient of filtration, and $B = \Delta f_{if}/\Delta f_{s.het}$ can be defined as the coefficient of utilization of the band.

Since we are concerned with the ratio $(P_s/P_n)_{out}$ for sufficiently high values of $(P_s/P_n)_{in}$ [usually for radio-relay lines $(P_s/P_n)_{in} > 30$ dB for peak ratios], Eq. (1) can be simplified. We obtain

$$\left(\frac{P_s}{P_n}\right)_{out} = 0.375\left(\frac{P_s}{P_n}\right)_{in}\frac{D^3}{B^2}.$$

(3)

It can be seen from Eq. (4) that the most effective method of increasing the signal-to-noise ratio at the input of the heterodyne frequency multiplier at PP is narrowing the transmission band ΔF of the low-frequency part of the FM receiver in the frequency channel 558-608 kHz. Using Eq. (4), depending on ΔF, ratios $(P_s/P_n)_{out}$ were calculated for two values of $\Delta f_{s.het}$ for a ratio $(P_s/P_n)_{in} = 30$ dB, corresponding to the worst conditions on the path R = 45 km. The relationships obtained are illustrated in Fig. 6. If, therefore, the necessary ratio $(U_{het}/U_{n\,het})_{in}$ is taken with a certain margin of 135 dB, the transmission band of the low-frequency part of the receiver must, according to Fig. 2, be $\Delta F \leq 100$ Hz.

The only practicable system for a retuneable filter with such a transmission band in frequencies 500-600 kHz is the system of automatic phase control (APC). In practice there is no difficulty in constructing an APC system with noise suppression of 135-140 dB. The phase error and slow phase drifts of the synchronized generator do not exceed 4-5° even after multiplication by 60 times. Consequently, the APC system is the best alternative for the formation of a coherent heterodyne voltage for the radio-astronomical receiver at PP. After leaving the radio-astronomical receiver at PP, the noise signal enters the delaying unit.

The delaying unit is so designed that, with a change in the delay time τ_d from 0 to 40 μsec, the transmission factor of the unit remains equal to unity. This is achieved because, when τ_d reaches its maximum, attenuation of the signal is compensated for by a special amplifier, so that the resultant transmission factor of the delaying unit is K = 1. Any detachable component of the delay line is substituted by a square attenuator, with attenuation equal to the attenuation of the signal in the omitted component.

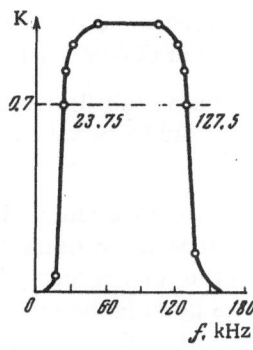

Fig. 8. Frequency characteristics of radio-astronomical receivers at output of balancing and filter unit.

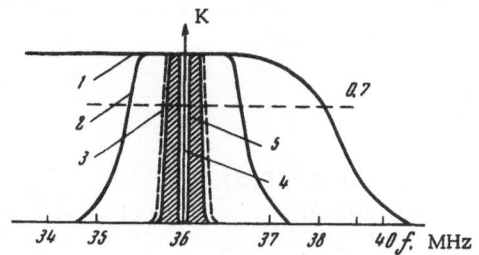

Fig. 9. Order of formation of frequency characteristic curve of radio-astronomical receiver at principal point. 1) Frequency characteristic (FC) of antenna amplifier; 2) FC of uhf-1; 3) FC of uhf-2; 4) heterodyne frequency; 5) FC of filter unit.

The frequency characteristics, their relationship, and the relationship between K and τ_d are shown in Fig. 7. Through the use of phase correctors, the phase characteristic curve for any value of τ_d within the limits of the working frequency band diverges from linear by not more than 0.3°. The delay line consists of a series of standard LC units giving delays of 5 and 1 μsec. A steady change in τ_d from 0 to 1 μsec is accomplished by a delay line working on the principle of a capacitive current collector [7]. This unit enables work to be carried out with bases up to 10–12 km. In the future, for longer bases, it is proposed to use a magnetostrictive L3. The noise signal at PP from L3 and the noise signal at DP from the radio-relay line receiver are fed into the balancing and filter unit.

In the balancing and filter unit the transmission factor in each of the two channels can be changed by up to 5 dB. Equality of signal levels at the output of the unit is regulated by a special differential voltmeter with an accuracy of 0.001 dB. The frequency characteristics of the channels are finally formed in this unit. For these purposes, a unit of interchangeable lower- and higher-frequency filters are used. Since the transmission bands of all previous instruments for noise signals from sources of radio waves are much wider than the transmission bands of the filter unit, the effective band and shape of the frequency characteristic curve of each channel are determined entirely by the filter unit. A pair of three-component filters of lower frequencies with f_{mean} = 120 kHz and a pair of three-component filters of higher frequencies with f_{mean} = 25 kHz are used for this purpose.

In this way high identity and stability of the transmission bands are obtained in both channels irrespective of changes in the parameters and working conditions of all preceding units.

Because of the interchangeability of the filters, the transmission bands of the channels can be varied without any need to retune the instruments. All the filters are constructed on standard interchangeable sections from the "Ural" computer. The transfer functions of the channels are illustrated in Fig. 8. The order of formation of the frequency characteristic curve of the PP channel is shown in Fig. 9.

The balancing and filter unit has two pairs of outputs. From one pair of outputs the noise signals are fed into the lf unit where they are summated, squared, detected, and, after passing through a narrow-band RC amplifier, are fed into a phase detector. Phase reversal with a frequency of 30 Hz is carried out at the heterodyne frequency.

The characteristic curve of the squaring circuit has a quadratic portion with an accuracy of 0.5% within limits of 1.5 V at the input of the summing cascade (7 V at the input of the squarer).

The other pair of outputs of the filter unit are connected to a panorama unit, providing for continuous visual control within the limits of the working band of each channel simultaneously.

The apparatus at each point includes a power supply with electronic stabilizers of anode and filament voltages. It also houses control units and alarm units for faults in all the principal components of the apparatus.

At all points, the signal received from radio-sources is checked by recording with the compensation method. At a given time the apparatus goes through a stage of combined adjustment and testing. The tests carried out enable the demands made on the apparatus to be specified more accurately in the various parameters described above. The good measure of agreement obtained between the experimental data and the calculated values gives hope that observations of sources of radio waves can be made with bases of up to 50 km.

In conclusion, the author wishes to thank V. V. Vitkevich for attention given to this work and for concrete suggestions made in the choice, discussion, and preparation of the apparatus for operation. The author is also grateful to M. T. Rezepin, V. A. Frolov, and to the staff of the Department of Radio Engineering of the Tula Polytechnic Institute for great help given during the designing and preparation of the instrumental part of the interferometer.

LITERATURE CITED

1. V. V. Vitkevich, Astron. Zh., Vol. 29, No. 4 (1952).
2. V. V. Vitkevich and R. S. Sorochenko, Astron. Zh., Vol. 30, No. 6 (1953).
3. R. N. Brown and R. Q. Twiss, Phil. Mag., Vol. 45, No. 336 (1954).
4. O. Elgaroy, D. Morris, and B. Rowson, Monthly Notices Vol. 124, No. 5 (1962).
5. J. Fago and F. Mann, Frequency Modulation in Radio Relay Lines [Russian translation], Sovetskoe Radio (1964).
6. E. S. Elinson and A. S. Larionov, Radiotekhnika, Vol. 22, No. 2 (1967).
7. A. G. Golubkov, Izvest. Vuz. Radiotekhn., Vol. 5, No. 4 (1962).